T0192457

AN INTRODUCTION TO COMBUSTION

Combustion Science and Technology Book Series

Editor
WILLIAM A. SIRIGNANO
University of California, Irvine, USA
Department of Mechanical and Aerospace Engineering

This book is part of a series. The publisher will accept continuation orders which may be cancelled at any time and which provide for automatic billing and shipping of each title in the series upon publication. Please write for details.

AN INTRODUCTION TO COMBUSTION

Warren C. Strahle

School of Aerospace Engineering
Georgia Institute of Technology
Atlanta, USA

CRC Press
Taylor & Francis Group
Boca Raton London New York

CRC Press is an imprint of the
Taylor & Francis Group, an **informa** business

First published 1993 by Overseas Publishers Association

Published 2019 by CRC Press
Taylor & Francis Group
6000 Broken Sound Parkway NW, Suite 300
Boca Raton, FL 33487-2742

© 1993 by Taylor & Francis Group, LLC
CRC Press is an imprint of Taylor & Francis Group, an Informa business

No claim to original U.S. Government works

ISBN 13: 978-2-88124-608-1 (pbk)

Visit the Taylor & Francis Web site at
http://www.taylorandfrancis.com

and the CRC Press Web site at
http://www.crcpress.com

Library of Congress Cataloging-in-Publication Data

Strahle, Warren C.
    An introduction to combustion / Warren C. Strahle.
    p.   cm. — (Combustion science and technology book series ; v.
1)
    Includes index.
    ISBN 2-88124-608-7
    1. Combustion. I. Title. II. Series.
    QD516.S886 1993
    541.3'61–dc20   92-42248
                                                 CIP

To my Mother, Josephine Marrs
for her years of support and understanding

CONTENTS

PREFACE

After nearly twenty-five years of teaching combustion-related material to university students, I have often noted that the subject is not easily grasped by undergraduate students in engineering. While there may be several reasons why this is so, a primary reason appears to be that they encounter the subject matter late in their undergraduate studies. Usually, the students are in their late junior or early senior years and the study of requisite chemistry and physics material is a distant memory. Moreover, combustion is a difficult subject, being truly interdisciplinary and requiring the merging of knowledge in several fields. While there are several outstanding combustion textbooks in print,[1] they are not suitable for introducing the subject at the undergraduate level. These texts are aimed primarily at an audience of graduate students and those with research interests in combustion. This text is written to appeal to the undergraduate engineering student who wishes an introduction to the subject, but not an exhaustive treatment.

Many of the technologically useful results in combustion are derived more by dimensional analysis and physical reasoning than by detailed mathematical analysis. Therefore, I have not emphasized mathematical developments. The main results and concepts are still physically correct and useful. As an example, when flammability limits due to heat loss are treated, an intuitive appeal is introduced that the limit occurs when the flame temperature drops below the ignition

---

[1]Some of these are: I. Glassman, *Combustion*. Orlando: Academic Press, 1987.
K. K. Kuo, *Principles of Combustion*. New York: Wiley-Interscience, 1986.
F. A. Williams, *Combustion Theory*. Redwood City: Addison-Wesley, 1985.

temperature. This is highly imprecise in most instances, but the concept is easily grasped.

I have tried to introduce most subjects by specific example, rather than by general development. It is my experience that this appeals to undergraduate students. They are generally more interested in the application of results rather than understanding why the result is obtained. Specific application examples are also included as an aid in problem solving. Problems appear at the end of each chapter with some solutions given. References to original literature and other sources are kept to a minimum, since it is my experience that the undergraduates do not use them. There are other texts that serve as an excellent source for further reading, but at a more advanced level.

A source of difficulty in the study of combustion is the wide variety of nomenclature in use because of the large number of disciplines involved. I have tried to use accepted nomenclature uniformly throughout and keep the introduction of new symbols to a minimum. The SI system of units is used as much as possible in this text even though the calorie permeates the combustion literature as an energy unit. Since the JANNAF Thermochemical Tables are now printed with the joule as the energy unit, it was deemed time for this adoption.

The book is suitable for a three credit hour, one semester course or, with some omissions, for a three or four hour, one quarter course. I sincerely hope that this book will fill an obvious void in the undergraduate textbook area.

ACKNOWLEDGMENTS

This book has been an undertaking to which many people have contributed. It actually began as an experimental program with an undergraduate class in the winter of 1992 at the Georgia Institute of Technology, AE 4261—Introduction to Aerospace Combustion. For each lecture I would write a set of notes or draw rough figures for typesetting by the students in the class. They had access to MacIntosh equipment and appropriate word processing, mathematics and drawing software in the undergraduate computer laboratory. In my office, however, I was tied to DOS-based equipment with little graphics software, so I let the students to the "dirty" work. I was also curious as to the educational benefit of asking the students to do the setup. At the end of the quarter, about 60% of the task had been completed, and I asked the students to comment on the usual confidential faculty evaluation forms concerning whether they had felt unnecessarily used for clerical help or whether there was any learning process involved in the task. Of the nearly forty students the response was overwhelmingly positive (95%) that, at least for the sections they were involved with, they learned a great deal from being confronted with the written and graphic material. I recommend the process to others who may feel that a text is needed for a particular course.

In any event, these students must be acknowledged. A list of the contributors follows:

Bryan E. Acree   David L. Buckwalter
Jeffrey R. Calcaterra   George W. Channel
J. Scott Culpepper   Suzette Cuneo
Donald H. Denton III   Samir M. Desai

William F. Forcey	Louis D. Gehrig
Troy D. Goodsen	Jack C. Griffis III
Charles E. Johnson	Renee M. Koch
Steven R. Luczynski	John C. Lynn
Ampei Morimoto	Patricia A. Morrison
Harry J. Nichols	Todd W. Parrish
Marc R. Robert	Charles E. Rolfes
Alice J. Savic	David M. Simon
Charles C. Somers III	Christopher T. Thornton
Maria K. Toogood	Matt B. Tucker III
Michael J. Weiland	Anna C. Wilson
James E. Wynn III	Thomas R. Young

Volunteering without pay, Mr. John C. Lynn did some special figure work after the quarter was over and his efforts are acknowledged. Some graduate students also helped during a graduate course (Combustion II) in the spring of 1992. These students are:

Steven J. Aarnio	Arindam Bose
Mark E. Burns	S. R. Chakravarthy
Emmett J. Davis	Douglas J. Declue

Chapter 8 was actually taken from some notes developed by a valued colleague, Dr. Jechiel I. Jagoda, who stepped in for me in 1990, during my recovery following medical treatment. Several conversations with him over the years have contributed greatly to this manuscript. Dr. Ben T. Zinn has also helped with my understanding of various areas since we have been close colleagues for twenty-four years.

Dr. William A. Sirignano, the series editor and a close personal friend, invited me to undertake this task (I don't know whether or not to thank him). Some editorial work was provided by a friend, Dr. Doris B. Fisher. For getting me interested in this subject and nurturing my knowledge base, there are four former teachers who deserve some credit. They are Drs. Robert Eustis and Walter Vincenti of Stanford University, and Drs. Irvin Glassman and the late Luigi Crocco of Princeton University.

Finally, and most important to the production of this book, Mrs. Terry Parrott, my secretary, is graciously thanked for being the disk manager, manuscript corrector and the helper of students in the typesetting experience. Her pleasant disposition in the face of some of my disorganization also helped keep this project on track.

ILLUSTRATIONS

FIGURES

NOMENCLATURE

There is uniform nomenclature throughout this book. Major nomenclature, which may include subscripts, will be defined below. When a subscript or superscript is used infrequently, it is defined in the separate subscript/superscript sections.

a	speed of sound, droplet radius
a_1, a_2	constants of integration
a_i	speed of sound at station i
A	pre-exponential factor in the Arrhenius law, area
A_s	surface area
c	concentration, speed of light
c_i	concentration of species i
c_p	specific heat at constant pressure (mass basis)
C	capacitance
C_p	specific heat at constant pressure (molar basis)
c_v	specific heat at constant volume (mass basis)
C_v	specific heat at constant volume (molar basis)
d	droplet diameter, radial or distance from symmetry axis
d_p	quenching distance
D	mass diffusion coefficient
e	internal energy per unit mass, radiated energy per molecule
E	internal energy per mole, activation energy
$\hat{E}$	extensive internal energy
f	ratio of mass of fuel to mass of oxidizer
f_s	stoichiometric ratio of mass of fuel to mass of oxidizer
G	Gibbs' free energy

h	enthalpy (mass basis), Planck's constant
H	enthalpy (molar basis)
H^0	molar basis enthalpy at standard state
$(\Delta H_f^0)_{T,i}$	molar basis heat of formation of species i at T and at standard state
$\hat{H}$	extensive enthalpy
I	radiant intensity
j	stoichiometric coefficient
k	specific reaction rate constant
k_f, k_b	forward and backward reaction rate constants
K_p	equilibrium constant based upon partial pressures
L	latent heat (enthalpy) of vaporization or sublimation
Le	Lewis number
L_f	flame length
$\dot{m}$	mass flow rate per unit area or sometimes total mass flow
M	total number of species
M_i	Mach number at station i
n	number of moles in a fixed volume system
n_i	number of moles of species i in a fixed mass system
N	number of molecules in a fixed mass system
N_0	Avogadro's number
N_i	number of molecules of species i in a fixed mass system
p	pressure
p_i	partial pressure of species i
Pe	Peclet number
Pr	Prandtl number
q_h	heat transfer rate per unit area
q	heat release per unit mass
$\dot{q}$	volumetric heat release rate
Q	heat release (molar basis)
Q_v	heat added at constant volume (molar basis)
Q_p	heat added at constant pressure (molar basis)
r	deflagration rate, distance from axis of symmetry
Re	Reynolds number
R	universal gas constant
R	reaction rate
$\mathbf{R}_{av}$	average reaction rate
s	entropy (mass basis)
S	entropy (molar basis)

$\hat{S}$	extensive entropy
S_L	laminar flame speed
T	absolute thermodynamic temperature
u	mass weighted velocity in major flow direction
u_i	absolute flow velocity of species i in major flow direction
u_{bo}	blowoff velocity
v	mass weighted velocity perpendicular to u, voltage
v_i	absolute velocity of species i perpendicular to major flow direction
V_i	diffusion velocity of species i
V	volume
w_i	volumetric production rate of species i by chemical reaction
w_F	volumetric mass production rate of fuel
W	molecular weight of mixture
W_i	molecular weight of species i
x	cartesian coordinate
X_i	mole fraction of species i
y	cartesian coordinate
Y_i	mass fraction of species i
z	grouping of stoichiometric coefficients
Z	dimensionless mass fraction variable grouping

GREEK NOTATION

α	thermal diffusivity
β	evaporation constant
γ	ratio of specific heats
δ	flame thickness, dimensionless ignition parameter
δ_c	thickness of conduction zone
δ_r	thickness of reaction zone
η	dimensionless distance variable, critical ignition parameter, critical value of the ratio of heat loss rate in a flame to heat generation rate in the flame
θ	half angle of laser beam intersection, dimensionless temperature variable
λ	coefficient of thermal conductivity
μ	coefficient of viscosity
ξ	extent of reaction
ρ	density

ρ_i	density of species i
σ	scattering cross section
τ_h	heat transfer time
τ_r	reaction time
Φ	equivalence ratio
Ω	optical constant

SUBSCRIPTS

b	burned, reverse action
c	conduction zone
f	formation reaction, forward reaction
F	fuel
h	heat transfer
i	species, direction, station number
ig	ignition
j	species
L	Laminar
O	oxidizer
p	constant pressure
pr	products
r	reaction
ref	reference state
s	stoichiometric, condensed phase–gas interface
T	temperature at which function is evaluated
u	unburned
v	constant volume
wb	wet bulb

SUPERSCRIPTS

$\rightarrow$	vector
$\wedge$	extensive quantity
0	standard state

INTRODUCTION
AND REVIEW

1.1. GENERAL INTRODUCTION

The generation of heat on earth has come largely from a process called combustion, and this will continue well into the next century. The process involves a chemical transformation between a substance or substances called *fuels* and other chemicals called *oxidizers*. No alteration of the nuclei of the entering substances is involved, but bonds involving the electrons of the molecules and atoms take place. This bonding causes heat to be liberated, or, in some cases, heat is required to form the bonds. The case of heat liberation is the most interesting, since it is this heat energy that can be usefully exploited. In addition to fuel and oxidizer, often an *ignition* source is needed. For example, one of the earliest forms of combustion, the forest fire, uses wood as the fuel, the air as the oxidizer and the ignition source can be a spark (lightning).[1]

Combustion has a wide variety of uses. It is used for power generation or creation of thrust in engines of all types. Combustion is used as a heat source for chemical processing, general heating and drying operations. It is useful in waste incineration, melting operations and welding. Combustion is involved in explosions for both useful and hostile purposes. The use of combustion has harmful effects, however. Pollutants and greenhouse gases are produced as well as waste heat and unwanted explosion and fires.

The science and engineering of combustion is the subject of this book. Combustion is a complex subject to master, requiring knowledge of many sub-disciplines of physics, chemistry and physical chemistry. Examples are fluid mechanics, quantum mechanics, statistical mechanics, kinetic theory, chemical kinetics, electromagnetic radiation, and heat and mass transfer. Fortunately, a working engineering knowledge can be attained without going too far into the details of these disciplines, and such an approach will be followed in this book.

Combustion can involve all phases of matter—solid, liquid and gas. For example, both solids and liquids are used in rocket engines, and gases are used in an oxy-acetylene welding torch. Although most of the material covered in this book will deal with the gas phase, the solid phase is covered, for example, in connection with solid rocket propellants and the liquid phase is dealt with when droplet burning is considered.

It is presumed, insofar as this book is concerned, that the student is at least a late junior in engineering or physics. Assumed is that first year physics, chemistry and calculus are past history and that first courses in thermodynamics and compressible fluid flow have been taken. However, with regard to compressible fluid flow, the necessary concepts are developed in this book, so that lack of knowledge here is perhaps not too crucial. The chemistry needed will be developed as necessary, but it is presumed that the student is familiar with the notions of atoms and molecules. Welcome to the study of combustion; it is a fascinating field.

---

[1]For an interesting history of man's use of combustion see F. J. Weinberg, *Fifteenth Symposium (International) on Combustion*. Pittsburgh: The Combustion Institute, 1974, pp. 1–20.

1.2. CHEMISTRY AND PHYSICS REVIEW

We shall often be concerned about the *atomic weight* and *molecular weight* of atoms and molecules. These weights are quoted relative to the carbon 12 atom. The number 12 is the number of particles in the nucleus, consisting, in this case, of six positively charged protons and six neutral neutrons. Carbon 12, being electrically neutral, has six negatively charged electrons, giving it the *atomic number* 6. The atomic weight or molecular weight of any other atom or molecule is twelve times the ratio of its actual weight to the weight of the carbon 12 atom. It should be clear that we are really considering masses, not weights. However, the literature is full of the term "weight," so by convention we adopt it here.

The proton and neutron have almost, but not exactly, the same mass and the electron has a nearly negligible mass compared to the proton. The consequence is that almost any atom or molecule will have its atomic or molecular weight very nearly an integer. As an example, the oxygen 16 atom has an atomic weight of 15.995, which is 16 for practical purposes.

There are also naturally occurring *isotopes* of the elements which have differing nucleus makeup from the most commonly occurring form of the element. For example, carbon 13 makes up about one percent of the naturally occurring carbon, with an extra neutron in the nucleus. As a consequence, the average atomic weight of carbon is 12.011, which, again, is 12 for practical purposes, an integer.

A common set of elements occurring in combustion problems is the C–H–N–O system. These elements have the atomic weights 12, 1, 14 and 16, respectively. The molecular weight of any compound is merely the sum of the weights of the atomic constituents. For example, the molecular weight of water, H_2O, is $1 + 1 + 16 = 18$. Similarly, the molecular weight of carbon dioxide, CO_2, is 44.

The next concept to review is that of the *mole* or, as sometimes abbreviated, the *mol*. It is defined as the mass in grams (g) of an element or compound equal to its atomic or molecular weight. For example, 12 g of carbon is equal to one mole of carbon; 44 g of CO_2 is one mole of the compound. It is clear, since the mole and the mass are synonymous, one mole of a compound or element consists of a fixed number of particles of the substance. This number is *Avogadro's* number, given the symbol N_0, and having the value 6.023×10^{23} particles/mole.

In the introduction of the mole, the unit of mass is the gram, and whenever the word "mole" appears alone the gram is understood to be the appropriate unit of mass. In the literature, however, other units of mass are often used. For example, the pound (lb) may be used so that 12 lb of carbon would correspond to 1 lb-mole, or 12 kg of carbon would be equivalent to 1 kg-mole of carbon. If any unit of mass other than the gram is used, it is imperative to state that unit when quoting the number of moles, because the amount of the substance differs between unit systems. Avogadro's number also differs from the above value if the mass unit differs from the gram.

Example 1.1.
What is Avogadro's number if the mass unit is the lb?

Solution.

$$N_0 = 6.023 \times 10^{23} \frac{\text{particles}}{\text{mole}} \times \frac{454 \text{ g}}{\text{lb}} = 2.734 \times 10^{26} \frac{\text{particles}}{\text{lb-mole}}.$$

We now wish to review the *mole fraction* and the *concentration*. Consider Fig. 1.2a where a container of volume V contains several molecules or atoms, therein separated by type, labeled 1, 2 or 3. Let the number of molecules of each species be denoted by N_i, where i is the species identifier. For example, $N_2 = 2$. The *concentration of species* i, c_i, is defined as $N_i/(N_0 V)$ and has the units of moles per unit volume. The overall concentration is obtained by merely summing up the concentrations of the individual components. That is,

$$c = \sum_{i=1}^{M} c_i = \frac{N}{N_0 V} = \frac{1}{N_0 V} \sum_{i=1}^{M} N_i \tag{1.1}$$

where M is the total number of species in V. In developing Eq. (1.1) we use the fact that any variable not depending on the subscript i may be removed from under the summation symbol. In Fig. 1.2a, $M = 3$.

In view of the large magnitude of N_0, it should be clear that even very small V can contain an enormous number of particles. In fact, V can be shrunk so low that it appears to an observer and to the calculus as a mathematical point, while still containing many particles. This is the *continuum* limit, where we may speak of concentration at a point, and it may vary from point to point in a flow, for example.

The *mole fraction of species* i, X_i, is defined as c_i/c which may vary from point to point in a system. Alternatively, in a macroscopic

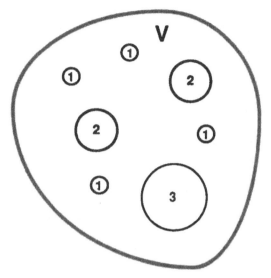

FIGURE 1.2a. A volume containing three molecule types.

system such as in Fig. 1.2a, $X_i = N_i/N$ where

$$N = \sum_{i=1}^{M} N_i.$$

Addition of all of the mole fractions must obviously equal unity. That is,

$$\sum_{i=1}^{M} X_i = \frac{1}{c} \sum_{i=1}^{M} c_i = \frac{c}{c} = 1;$$

consequently only $M - 1$ of the mole fractions are independent. Specification of $M - 1$ of them specifies the last.

The molecular weight of species i will be given by the symbol W_i. From it and the concentration of species i we may construct the *density of species* i, or the mass per unit volume of species i, as $\rho_i = c_i W_i$. The overall density, ρ, follows as

$$\rho = \sum_{i=1}^{M} \rho_i.$$

The overall molecular weight of a mixture, the mass per unit mole, then follows as

$$W = \frac{\text{mass}}{\text{mole}} = \frac{\sum_{i=1}^{M} c_i W_i}{c} = \sum_{i=1}^{M} \frac{c_i}{c} W_i = \sum_{i=1}^{M} X_i W_i. \qquad (1.2)$$

The *mass fraction of species* i is formed in a fashion similar to the mole fraction and is defined as $Y_i = \rho_i/\rho$. The relation between the mole fraction and the mass fraction then becomes

$$Y_i = \frac{c_i W_i}{\sum_{i=1}^{M} c_i W_i} = \frac{c_i W_i/c}{\sum_{i=1}^{M} c_i W_i/c} = \frac{X_i W_i}{W}. \qquad (1.3)$$

When we manipulate Eq. (1.3), another method of calculating mixture molecular weight appears. That is,

$$\sum_{i=1}^{M} \frac{Y_i}{W_i} = \sum_{i=1}^{M} \frac{X_i}{W} = \frac{1}{W} \sum_{i=1}^{M} X_i = \frac{1}{W}. \qquad (1.4)$$

Example 1.2.
In a vessel of volume 1 m³ there are 1 g of molecular hydrogen, H_2, and 3 g of molecular oxygen, O_2. What are the mole fractions, mass fractions and concentrations of the oxygen and hydrogen? What are the overall density and mixture molecular weight?

Solution.

Total mass = $1 + 3 = 4$ g,
Moles H_2 = 1 g/(2g/mole) = 0.5 moles,
Moles O_2 = 3/(2 × 16) = .09375,
Total moles = 0.59375,
X_{H_2} = 0.5/0.59375 = 0.8421, X_{O_2} = .09375/0.59375 = 0.1579,
Y_{H_2} = 1/4 = 0.25, Y_{O_2} = 3/4 = 0.75,
c_{H_2} = 0.5 moles/1m³ = 0.5 moles/m³,
c_{O_2} = 0.09375/1 = 0.09375 moles/m³,
ρ = 4 g/1 m³ = 4 g/m³,
$W = \sum_i c_i W_i = 0.5 \cdot 2 + 0.09375 \cdot 32 = 4.0$ g/mol.

Finally, by way of review, we investigate the method of writing down chemical reactions and the law of stoichiometry. This law states that in the absence of nuclear reactions a chemical reaction must preserve the nuclei and the total number of electrons. For example, a possible reaction is

$$(1)C + (1)O_2 \rightarrow (1)CO_2.$$

This is to be read as "one molecule (or mole) of carbon plus one molecule (or mole) of molecular oxygen goes to one molecule (or mole) of carbon dioxide." The numbers multiplying the species are called *stoichiometric coefficients*. When the number is unity it is always omitted. The above reaction would actually be written as

$$C + O_2 \rightarrow CO_2.$$

Another example is

$$2C + O_2 \rightarrow 2CO.$$

An example of a reaction stripping electrons, called an *ionization reaction*, is

$$2NO \rightarrow NO + NO^+ + e^-.$$

The key to the above is that the total number of letter symbols, which signify the nucleus composition, must be conserved in going from left to right. The overall charge must also be conserved.

Insofar as the law of stoichiometry is concerned, there is no violation if the above reactions proceed from right to left as well as left to right. If this is indeed occurring we write the reaction as, for example,

$$2C + O_2 \rightleftarrows 2CO.$$

If the reaction only proceeds one way (left to right, for example) we call the species on the left the *reactants* and those on the right the *products*. This distinction is lost if the reaction proceeds both ways.

Finally, we will sometimes have reactions between multiple phases, and we must distinguish the phases in these cases. If solid carbon, for example, reacts with liquid molecular oxygen to form gaseous carbon monoxide, we may write

$$2C(s) + O_2(l) \rightarrow 2CO(g)$$

where s, l and g stand for solid, liquid and gas, respectively. If there is no indication of the phase, it is presumed that only gas phase species are present.

1.3. PROBLEMS

1. Write the stoichiometric coefficients for the products in the following reactions:

   a. $H_2 + 3O_2 \rightarrow ?H_2O + ?O_2$
   b. $H_2 + \frac{1}{2}F_2 \rightarrow ?HF + ?H_2$
   c. $CH_4 + 3O_2 \rightarrow ?CO_2 + ?H_2O + ?O_2$
   d. $C_2H_5OH + 6O_2 \rightarrow ?CO_2 + ?H_2O + ?O_2$

2. In a steady flow system 1 mol/s of CH_4 and 2 mol/s of O_2 enter a reactor and become H_2O and CO_2 exiting the reactor. The density (gas) at the exit is 1 kg/m^3. What is the density of the water vapor at the exit? What is the concentration of CO_2 at the exit?

2

CHEMICAL THERMODYNAMICS

2.1. INTRODUCTION

Thermodynamics is an empirical science dealing with the properties of substances and their energetics. It applies to systems of a single phase and multiple phases. For purposes here it also applies to systems of varying chemical composition. The subject of chemical thermodynamics applies to fixed mass and flowing systems, just as in the case of fixed composition systems. The subject simply becomes a bit more complex when chemical change is involved, than in the case of fixed composition systems. Since combustion involves composition changes, however, this complexity must be accepted as necessary.

9

Thermodynamics enables us to calculate the energetics of system changes in composition. As such it enables us to determine, for example, the temperature and pressure changes when a system undergoes a chemical transformation. It will be seen that thermodynamics can also be used to tell us what the composition change will be when a system undergoes a reaction. It is not used, however, to determine rates of chemical transition. That is the subject of a following chapter on chemical kinetics. The subject of thermodynamics is only concerned with beginning and end thermodynamic states for a system, with no concern for the process path between them. Nevertheless, it is an essential science in combustion.

2.2. PROPERTIES OF SUBSTANCES

We shall deal with solids, liquids and gases. For solids and liquids we will use the actual measured thermodynamic properties of the substances. So, for example, any thermodynamic equation of state such as a relation between pressure, temperature and volume for liquids and solids will be the actual measured relation. For gases, however, but with good approximation, we will always assume them to be *thermally perfect gases*. The term "thermally perfect" means that the gas obeys the equation of state

$$pV = nRT \qquad (2.1)$$

where V is a fixed system volume, p is pressure and T is absolute thermodynamic temperature. The symbol n is the number of moles in a fixed mass system given by

$$n = \sum_{i=1}^{M}(N_i/N_0) = \sum_{i=1}^{M} n_i.$$

The symbol R is the *universal gas constant* given numerically as

$$R = 8314\frac{J}{kmol \cdot K} = 8.314\frac{J}{mol \cdot K}$$

$$= 1554\frac{ft \ \#}{lb\text{-}mol \cdot R} = 1.987\frac{cal}{mol \cdot K}$$

where the # symbol is pounds-force, 1 lb accelerated at the rate of 32.2 ft/s^2. The energy unit "calorie" (cal) is not the currently approved energy unit in the international system of units (SI system). However, it permeates the combustion literature so heavily and is so often used in tables of thermodynamic properties that the reader should be aware of it. The conversion between the calorie and Joule is

$$1 \text{ cal} = 4.186 \text{ J.}$$

Dividing Eq. (2.1) by V, we obtain

$$p = cRT = \rho RT/W$$

which is now in terms of variables which are independent of system size and may vary from point to point in a system. Such variables are called *intensive* thermodynamic state variables while those depending upon system size, such as V, are called *extensive* variables.

Thermally perfect gases have the property that if a species other than a given species occupying a container is added to the container the normal force (pressure) of the original species acting on the container walls is not altered. In a mixture of several different gases the pressure on the wall due to each species i is p_i. Each of these *partial pressures* may be computed by

$$p_i = c_i RT \tag{2.2}$$

and each p_i is independent of the others. It follows that the pressure from all of the species may be computed by

$$p = \sum_{i=1}^{M} p_i = \sum_{i=1}^{M} c_i RT = RT \sum_{i=1}^{M} c_i = cRT$$

since RT does not depend upon the summation index i and may be taken from under the summation sign. This is known as Dalton's law of partial pressures.

Turning now to the variables describing various kinds of energy, the nomenclature to be used here is that small letters will describe variables on a per unit mass basis. So, for example, the internal energy on a mass basis will be given the symbol e. The enthalpy on a per unit mass basis is h; this is a defined variable where, by definition, $h \equiv e + p/\rho$ and is very useful in flow problems. If these variables are given on a molar basis (per unit mole), capital letters will be used, E and $H = E + p/c$. Similarly, the entropy will be denoted by S or s, depending upon the basis.

It is a very useful fact that gases which obey Eq. (2.1) have the property that e and h (and E and H) are functions only of temperature. In fact,

$$e = \int_{T_{ref}}^{T} c_v(T)\,dT + e_{ref} = e(T \text{ alone}) \qquad (2.3)$$

where the subscript "ref" denotes a reference state to be described later and $c_v = c_v(T \text{ alone})$ is the *specific heat at constant volume*. Similarly, for a given pure perfect gas substance

$$h = e + \frac{p}{\rho} = e(T) + \frac{R}{W}T = h(T) = \int_{T_{ref}}^{T} c_p(T)\,dT + h_{ref} \qquad (2.4)$$

where c_p is the *specific heat at constant pressure*. The specific heats on a molar basis are denoted by C_v and C_p, respectively. The conversion between the molar and mass basis is through the molecular weight. That is, $E = eW$ and $H = hW$.

In a mixture of species, whether solids, liquids and/or gases are present, we will always assume that there is no interaction between the species, so that the energies or enthalpies of the individual species are the same as if the species were alone. This has the consequence that the quantities evaluated for the entire mixture are merely the sums of the quantities for the individual species. That is,

$$E = \sum_{i=1}^{M} X_i E_i, \qquad H = \sum_{i=1}^{M} X_i H_i, \qquad e = \sum_{i=1}^{M} Y_i e_i, \qquad h = \sum_{i=1}^{M} Y_i h_i$$

where the E_i and H_i are the same as if the species were alone. The reader should be cautioned that this is not always the case. Care is sometimes required when there are inter-species interactions, but we shall not encounter this problem in this text.

When we deal with extensive quantities for systems of a finite size, we will use a "hat" superscript. Therefore, we have

$$\hat{E} = \sum_{i=1}^{M} n_i E_i, \qquad \hat{H} = \sum_{i=1}^{M} n_i H_i$$

as the energies for the complete system. The same conventions as above apply to the entropies, S, s and $\hat{S}$.

2.3. HEATS OF REACTION AND FORMATION

Consider the first law of thermodynamics applied to a fixed mass system, where the system undergoes a change of state from condition 1 to 2. It is presumed that changes in kinetic and potential energy are negligible. Then

$$\hat{E}_2 - \hat{E}_1 = \Delta \hat{E} = \text{heat added} - \text{work done by the system on the surroundings.}$$

This law is empirically valid even if the composition has changed in going from state 1 to state 2. Presume further that there is no work done. That is, there is no shaft work and the system remains at constant volume so that the pressure does no mechanical work on the surroundings. In this case

$$\Delta \hat{E} = Q_v \qquad (2.5)$$

where Q_v is the heat added at constant volume. An example would be an explosion in a constant volume vessel or a room if the walls or windows did not break. A second example of application of the first law is for a process taking place at constant pressure. Figure 2.3a schematically shows the initial state and final state of the system, consisting of a frictionless, perfectly sealed piston of weight w, resting upon the contents in the vessel. Here, work is performed on the surroundings if the contents of the vessel expand in going from state 1 to state 2. The first law applied here yields

$$\Delta \hat{E} = Q_p - p \cdot (V_2 - V_1) \quad \text{or} \quad \Delta \hat{H} = Q_p \qquad (2.6)$$

where Q_p denotes the heat added in a constant pressure process. It is seen that the energy is the natural variable in a constant volume process and the enthalpy is natural for constant pressure processes (hence, its usefulness by definition).

A constant volume chemical conversion can occur in a confined vessel, as mentioned. It may also be a good model for the combustion in an automobile cylinder, if the combustion takes place in a short enough time that the piston does not move too far. In steady flow rocket or jet engines the combustion takes place, usually, at nearly constant pressure, so the constant pressure assumption is closely valid. In any event, the reader should recall from thermodynamics that the enthalpy is the most natural variable for treating steady flow systems.

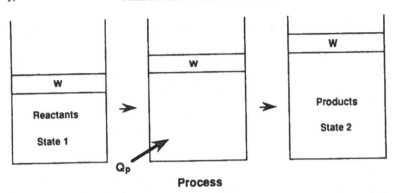

FIGURE 2.3a. Schematic of constant pressure process with heat addition.

Which variable is used, the energy or the enthalpy, is purely a matter of algebraic convenience, however.

Consider now the reaction of two *propellants* often considered for rocket engine use, hydrogen (H_2) and fluorine (F_2). Consider the reactants to be 1/2 mole hydrogen and 1/2 mole fluorine. The product is hydrogen fluoride (HF). The reaction is

$$\tfrac{1}{2}H_2 + \tfrac{1}{2}F_2 \to HF$$

in the case where all hydrogen and fluorine are consumed. Consider also that the pressure and temperature of the initial and final states are the same. How can this be so, given the fact that intuition tells us that this propellant combination is a powerful explosive and energy will be released? The situation is as shown in Fig. 2.3b. The pressure is maintained constant by the piston and the temperature can be manipulated by the extraction of heat. Thus, an overall constant pressure, constant temperature process can be forced, if the proper amount of heat is removed.

Because of its importance as a reference condition, assume that the pressure and temperature for this process are 1 bar (10^5 Pa, with 1 Pa = 1 N/m^2) and 298 K, respectively. Under these conditions the substances above behave as perfect gases (where from now on "perfect" will be used for "thermally perfect"). The enthalpies are therefore only functions of temperature and, for any species i, will be denoted as $(H_T)_i$. Applying Eq. (2.6) to this process, we have

$$Q_P = (H_{298})_{HF} - \tfrac{1}{2}(H_{298})_{H_2} - \tfrac{1}{2}(H_{298})_{F_2}.$$

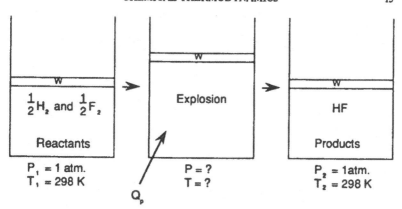

FIGURE 2.3b. Constant pressure, constant temperature combustion of hydrogen and fluorine.

If this experiment is actually carried out, the result is

$$Q_P = -272.55 \text{ kJ} = -64.2 \text{ kcal.}$$

The negative sign merely means that heat is evolved rather than added. Reactions of this kind are called *exothermic*; if heat were required to be added to keep the temperature the same, the reaction would be called *endothermic*.

The quantity $-Q_P$ for this reaction is called the *heat of reaction* at the *reference state*. The heat of reaction is positive for exothermic reactions. If the pressure and/or temperature were not at 1 bar and/or 298 K the $-Q_P$ would be the heat of reaction at the stated pressure and temperature, which must, in fact, be stated. The heat of reaction will differ from the above number, depending upon the thermodynamic conditions.

We now make some further definitions. For solids and liquids, the *standard state* of a substance is its real state at a pressure of 1 bar, for any temperature T. The standard state for a gas is its (real or fictitious) ideal gas state at 1 bar at any temperature T. To understand what is meant here, it should be emphasized that all gases have some pressure and temperature conditions under which they do not obey Eq. (2.2). Especially near the liquefaction condition, the perfect gas law simply does not hold. As an example, water vapor at the boiling point of liquid water is not a perfect gas. But we define the standard state of water vapor at 1 bar and 373 K (the boiling point) as the fictitious ideal (perfect) gas state at these conditions. While the

methods used to get numbers for such a fictitious state are beyond
the scope of this text, the methods are basically a theoretical extrap-
olation of thermodynamic properties to the standard state conditions
from conditions where water vapor does, in fact, behave perfectly. As
an example, at 1000 K and 1 bar. water vapor is, for all practical purposes, ideal and the thermodynamic properties may be extrapolated
to the boiling point assuming it to be ideal.

We denote the standard state with a superscript "0". In the above
hydrogen–fluorine example the substances, in fact, behave as ideal
gases so that we may write at the 1 bar, 298 K condition

$$Q_P = (H^0_{298})_{HF} - \tfrac{1}{2}(H^0_{298})_{H_2} - \tfrac{1}{2}(H^0_{298})_{F_2}.$$

The Q_P comes out as a difference between the enthalpies of the product and the elements from which it is formed.

In principle, all compounds can be formed by reactions by their
constituent elements. In thermodynamics we never know absolute values of energy variables, but that is not important because only changes
in these variables are dealt with in thermodynamics. To facilitate numerical work, we define a reference set of substances and their thermodynamic properties. We define reference chemicals as the elements
in the form most abundantly found in nature when they are found
alone. So, for example, oxygen in the air is found as the diatomic
molecule O_2 in the gas phase. The reference state for oxygen is the
gaseous diatomic molecule. Similarly, reference states for fluorine, hydrogen, and nitrogen are gaseous F_2, H_2, and N_2 respectively, not
F, H, and N. By contrast, the reference state for carbon is $C(s)$, or
graphite.

We define *formation* reactions for a compound or atom as the reaction that forms one mole of the substance from the elements or
element in their reference state. The *heat of formation* for a substance is the standard state enthalpy change for a reaction forming
the substance from its reference state elements. Moreover, for thermodynamic reference the heat of formation of the elements in their
standard states is defined as zero. The heat of formation may be defined at any temperature and for species i. It is denoted as

$$(\Delta H^0_f)_{T,i}.$$

In the above hydrogen–fluorine reaction it is clear that it was a formation reaction at 298 K, since hydrogen and fluorine behave perfectly

TABLE 2.3a
Heats of formation at 298 K

Chemical Symbol	Name	State	ΔH_f^0(kJ/mol)	Δh_f^0(kJ/g)
N	Nitrogen atom	Gas	472.66	33.760
O	Oxygen atom	Gas	249.19	15.577
H	Hydrogen atom	Gas	217.94	217.941
NO	Nitric oxide	Gas	90.37	3.012
OH	Hydroxyl radical	Gas	42.09	2.477
O_2	Oxygen	Gas	0	0
N_2	Nitrogen	Gas	0	0
H_2	Hydrogen	Gas	0	0
C	Carbon	Solid	0	0
CH_4	Methane	Gas	−74.85	−4.678
CO	Carbon monoxide	Gas	−109.79	−3.950
H_2O	Water	Gas	−241.83	−13.435
C_8H_{18}	Octane	Liquid	−249.95	−2.192
H_2O	Water	Liquid	−285.85	−15.882
CO_2	Carbon dioxide	Gas	−393.50	−8.945

at the stated conditions. Therefore,

$$(\Delta H_f^0)_{298,HF} = -272.6 \text{ kJ}.$$

Other examples of formation reactions are

$$H_2 + \tfrac{1}{2}O_2 \rightarrow H_2O(l), \qquad (\Delta H_f^0)_{298,H_2O(l)} = -286.0 \text{ kJ},$$

$$C(s) + O_2(g) \rightarrow CO_2(g), \qquad (\Delta H_f^0)_{298,CO_2(g)} = -393.5 \text{ kJ}.$$

A short list of heats of formation is given in Table 2.3a where both the molar form and the mass basis form are given. Notice that the heat of formation of atoms is positive, if the reference state of the atom is diatomic. This occurs because strong chemical bonds must be broken to form the atoms; the process is endothermic. Notice also, because it will be important later, that the compounds with the highest *negative* heat of formation in the C–H–O system are CO_2 and H_2O; the formation process for these compounds is exothermic.

An extensive set of thermodynamic properties of substances is provided by the Joint Army Navy NASA Air Force Thermochemical Tables (JANNAF Tables). Several of the constituent tables are provided in Appendix A of this text. One, for ideal gas water, is shown as Ta-

CHEMICAL THERMODYNAMICS

TABLE 2.3b
Thermodynamic properties of water from JANNAF tables
(Water (H_2O))
(Ideal gas)($W = 18.02$)
Enthalpy reference temperature $= T_r = 298.15$ K
Standard state pressure $= p^0 = 0.1$ MPa

| | J K^{-1}mol^{-1} | | | kJ mol^{-1} | | | |
T/K	C_p^0	S^0	$-[G^0-H^0(T_r)]/T$	$H^0-H^0(T_r)$	ΔH_f^0	ΔG_f^0	LogK_f
0	0.000	0.000	Infinite	−9.904	−238.921	−238.921	Infinite
100	33.299	152.388	218.534	−6.615	−240.083	−236.584	123.579
200	33.349	175.485	191.896	−3.282	−240.900	−232.766	60.792
298	33.590	188.834	188.834	0.000	−241.826	−228.582	40.047
300	33.596	189.042	188.835	0.062	−241.844	−228.500	39.785
400	34.262	198.788	190.159	3.452	−242.846	−223.901	29.238
500	35.226	206.534	192.685	6.925	−243.826	−219.051	22.884
600	36.325	213.052	195.550	10.501	−244.758	−214.007	18.631
700	37.495	218.739	198.465	10.501	−244.758	−214.007	18.631
800	38.721	223.825	201.322	18.002	−246.443	−203.496	13.287
900	39.987	228.459	204.084	21.938	−247.185	−198.083	11.496
1000	41.268	232.738	206.738	26.000	−247.857	−192.590	10.060
1100	42.536	236.731	209.285	30.191	−248.460	−187.033	8.881
1200	43.768	240.485	211.730	34.506	−248.997	−181.425	7.897
1300	44.945	244.035	214.080	38.942	−249.473	−175.774	7.063
1400	46.054	247.407	216.341	43.493	−249.894	−170.089	6.346
1500	47.090	250.620	218.520	48.151	−250.265	−164.376	5.724
1600	48.050	253.690	220.623	52.908	−250.592	−158.639	5.179
1700	48.935	256.630	222.655	57.758	−250.881	−152.883	4.698
1800	49.749	259.451	224.621	62.693	−251.138	−147.111	4.269
1900	50.496	262.161	226.526	67.706	−251.368	−141.325	3.885
2000	51.180	264.769	228.374	72.790	−251.575	−135.528	3.540
2100	51.823	267.282	230.167	77.941	−251.762	−129.721	3.227
2200	52.408	269.706	231.909	83.153	−251.934	−123.905	2.942
2300	52.947	272.048	233.604	88.421	−252.092	−118.082	2.682
2400	53.444	274.312	235.253	93.741	−252.239	−112.252	2.443
2500	53.904	276.503	236.860	99.108	−252.379	−106.416	2.223
2600	54.329	278.625	238.425	104.520	−252.513	−100.575	2.021
2700	54.723	280.683	239.952	109.973	−252.643	−94.729	1.833
2800	55.089	282.680	241.443	115.464	−252.771	−88.878	1.658
2900	55.430	284.619	242.899	120.990	−252.897	−83.023	1.495
3000	55.748	286.504	244.321	126.549	−253.024	−77.163	1.344

TABLE 2.3b
Continued

	J K^{-1}mol^{-1}			kJ mol^{-1}			
T/K	C_p^0	S^0	$-[G^0-H^0(T_r)]/T$	$H^0-H^0(T_r)$	ΔH_f^0	ΔG_f^0	Log K_f
3100	56.044	288.337	245.711	132.139	−253.152	−71.298	1.201
3200	56.323	290.120	247.071	137.757	−253.282	−65.430	1.068
3300	56.583	291.858	248.402	143.403	−253.416	−59.558	0.943
3400	56.828	293.550	249.705	149.073	−253.553	−53.681	0.825
3500	57.058	295.201	250.982	154.768	−253.696	−47.801	0.713
3600	57.276	296.812	252.233	160.485	−253.844	−41.916	0.608
3700	57.480	298.384	253.459	166.222	−253.997	−36.027	0.509
3800	57.675	299.919	254.661	171.980	−254.158	−30.133	0.414
3900	57.859	301.420	255.841	177.757	−254.326	−24.236	0.325
4000	58.033	302.887	256.999	183.552	−254.501	−18.334	0.239
4100	58.199	304.322	258.136	189.363	−254.684	−12.427	0.158
4200	58.357	305.726	259.252	195.191	−254.876	−6.516	0.081
4300	58.507	307.101	260.349	201.034	−255.078	−0.600	0.007
4400	58.650	308.448	261.427	206.892	−255.288	5.320	−0.063
4500	58.787	309.767	262.486	212.764	−255.508	11.245	−0.131
4600	58.918	311.061	263.528	218.650	−255.738	17.175	−0.195
4700	59.044	312.329	264.553	224.548	−255.978	23.111	−0.257
4800	59.164	313.574	265.562	230.458	−256.229	29.052	−0.316
4900	59.275	314.795	266.554	236.380	−256.491	34.998	−0.373
5000	59.390	315.993	267.531	242.313	−256.763	40.949	−0.428
5100	59.509	317.171	268.493	248.258	−257.046	46.906	−0.480
5200	59.628	318.327	269.440	254.215	−257.338	52.869	−0.531
5300	59.746	319.464	270.373	260.184	−257.639	58.838	−0.580
5400	59.864	320.582	271.293	266.164	−257.950	64.811	−0.627
5500	59.982	321.682	272.199	272.157	−258.268	70.791	−0.672
5600	60.100	322.764	273.092	278.161	−258.595	76.777	−0.716
5700	60.218	323.828	273.973	284.177	−258.930	82.769	−0.758
5800	60.335	324.877	274.841	290.204	−259.272	88.767	−0.799
5900	60.453	325.909	275.698	296.244	−259.621	94.770	−0.839
6000	60.571	326.926	276.544	302.295	−259.977	100.780	−0.877

Previous: March 1961 (1 atm) Current: March 1979 (1 bar)
Water (H$_2$O) H$_2$O$_1$(g)

ble 2.3b here. The first column is the temperature and the sixth column is the heat of formation at the corresponding temperature. If the thermodynamic properties of a reference substance were being tabulated, column six would show a zero at *any* temperature. For example, for H_2, the formation reaction would be the identity $H_2 \rightarrow H_2$ with a $Q_P = 0$.

The numbers for heats of formation are empirically derived; experiments must be performed to obtain them. Some of the numbers must be obtained by indirect means, using principles of thermodynamics. For example, consider the formation of CO:

$$C(s) + \tfrac{1}{2}O_2(g) \rightarrow CO(g).$$

This formation reaction cannot be carried out in the laboratory, because CO_2 is formed as well as CO. However, the formation reaction for CO_2 can be carried out with almost a perfect yield for the carbon dioxide. The reaction is given above with a heat of formation of -393.5 kJ/mol. It is also possible to form and store pure CO and to do the following reaction at the reference conditions:

$$CO + \tfrac{1}{2}O_2 \rightarrow CO_2.$$

The heat of reaction at 1 bar and 298 K is -283.0 kJ per mol of CO_2 formed. The reverse reaction must yield a heat of reaction of $+283.0$ kJ. The following reactions may be written, violating no law of thermodynamics or stoichiometry:

$$C(s) + \tfrac{1}{2}O_2 + \tfrac{1}{2}O_2 \rightarrow CO_2 \rightarrow CO + \tfrac{1}{2}O_2.$$

The overall process merely carries 1/2 mole O_2 along "for the ride" with no thermodynamic change. Consequently, the overall process is one of formation of one mole of CO from the elements and the heat of formation of CO must be

$$(\Delta H_f^0)_{298,CO} = -393.5 + 283.0 = -110.5 \text{ kJ/mol.}$$

Now consider a more complex process. Consider a mixture of 1/2 mole H_2 and 1/2 mole F_2 at 5 bar pressure and 200 K. We ask what happens if the reaction takes place at constant pressure to one mole HF and the mixture ends up at 2000 K. In particular, how much heat must be liberated or absorbed in this process? Thermodynamics does not care about the process path, and the result must be the same for any path as long as the end states are as specified. We will choose the path so that the calculations are convenient. Note, however, that the end states are no longer at the standard state and the temperatures are not the reference temperature of 298 K. Consider the rather

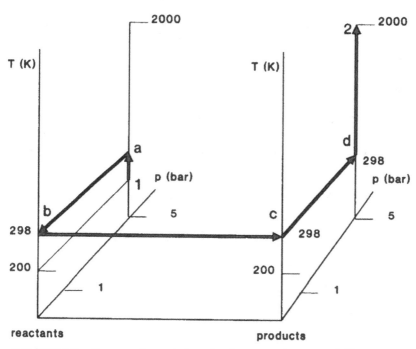

FIGURE 2.3c. Process schematic in going from state 1 to state 2. This is the complicated way.

complex path in Fig. 2.3c, where the reactants are on the left and the product is on the right. Shown is the following path:

$1 \to a$. Heat the reactants at constant pressure to the reference temperature of 298 K.

$a \to b$. At constant temperature drop the pressure to the standard state pressure of 1 bar.

$b \to c$. React the reactants to product at constant pressure and temperature.

$c \to d$. Compress the product to 5 bar at constant temperature.

$d \to 2$. Heat up the product to 2000 K at constant pressure.

Because the overall process is one at constant pressure, it is true that

$$\Delta \hat{H}_{1-2} = \hat{H}_2 - \hat{H}_1 = Q_p = n_2 HF - n_{\mathrm{H}_2,1} H_{1,\mathrm{H}_2} - n_{\mathrm{F}_2,1} H_{1,\mathrm{F}_2}$$

$$(2.7)$$

where the overall enthalpy change is that of the sum of the enthalpy changes for each step along the path. However, the decompression and compression processes of steps a → b and c → d require both heat transfer and work to be done on and by the fluid. In those processes the enthalpy change is not merely the heat added or subtracted.

Path 1 → a is a constant pressure heating process. The reactants are perfect gases. Looking at Table 2.3a, column 5 of the JANNAF Tables gives us the enthalpy change in going from 298 K to any other temperature. From Appendix A for hydrogen and fluorine, the reactants,

$$Q_{p,1-a} = \tfrac{1}{2}(H^0_{298,H_2} - H^0_{200,H_2} + H^0_{298,F_2} - H^0_{200,F_2}) = \tfrac{1}{2}(2.77 + 2.99).$$

Step b → c is a formation reaction at 1 bar and 298 K, for which

$$Q_{p,1-a} = (\Delta H^0_f)_{298,HF} = -272.55 \text{ kJ}.$$

Path d → 2 is a constant pressure heating process for pure HF, which from Appendix A requires an enthalpy change of

$$Q_{p,d-2} = (H^0_{2000,HF} - H^0_{298,HF}) = +52.83 \text{ kJ}.$$

Paths a → b and c → d have no enthalpy change associated because the enthalpy only depends upon temperature for perfect gases. Consequently, the overall enthalpy change in going from 1 to 2 is

$$\Delta \hat{H}_{1-2} = 52.83 - 272.55 + \tfrac{1}{2}(5.76) = -216.84 \text{ kJ} = Q_{p,1-2}.$$

Notice that this Q_p is negative, so that the overall process is exothermic, even in view of the fact that the temperature has increased. These reactants, if heat is not transferred out of the system, would go to an even higher temperature than 2000 K.

There is a simpler way of accomplishing the overall state change. Recalling that the enthalpy of a perfect gas independent of pressure, we see that the results will not depend upon the fact that we are at 5 bar. Figure 2.3d shows a simpler path for calculation. It is:

1 → a. Same as above.

a → b. Reaction at 5 bar and 298 K. This is a formation
         reaction for HF.

b → 2. Same as d → 2 above.

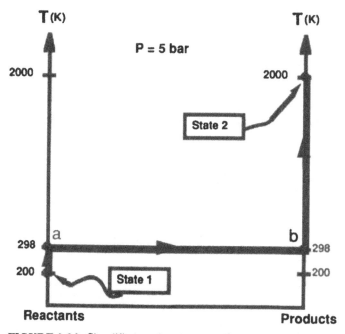

FIGURE 2.3d. Simplified perfect gas transition to the final state.

For each step the Q_p is the enthalpy change for that step and the overall Q_p is merely the sum of the heat added or subtracted at each step. That is,

$$Q_p = \tfrac{1}{2}(5.76) - 272.55 + 52.83 = -216.84 \text{ kJ}$$

which is the same as in the previous calculation.

To generalize the steps above, the enthalpy for any substance i is written as

$$H_{T,i} = H_{T,i} - H^0_{298,i} + (\Delta H^0_f)_{298,i} \qquad (2.8)$$

where the difference between the first two terms on the right side is called the *sensible enthalpy* and the last term is, of course, the heat of formation. This nomenclature is somewhat confusing, since Eq. (2.8) is an identity. The last two terms are, in fact, equal, but convention gives them different symbols. Shown here is the reference temperature of 298 K. Any other temperature could have been chosen, but

the JANNAF Tables make the use of 298 K most convenient. The
reader should verify that the placement of Eq. (2.8) in Eq. (2.7) will
reproduce the numerics of the example.

Example 2.1.
In the hydrogen–fluorine case above, calculate the heat added if the
process path follows (a) heating of the reactants from 200 K to 2000 K
at the pressure of 5 bar, and then (b) conversion to product at 5 bar.
and 2000 K.

Solution.
From the tables in Appendix A:

$$H_{200,H_2} = -2.77 + 0,$$

$$H_{2000,H_2} = 52.95 + 0,$$

$$H_{200,F_2} = -2.99 + 0,$$

$$H_{2000,F_2} = 62.75 + 0,$$

$$(\Delta H_f^0)_{2000,HF} = -277.57.$$

For process (a)

$$Q_{P,a} = \tfrac{1}{2}(52.95 + 2.77 + 62.78 + 2.99) = 60.75.$$

For process (b)

$$Q_{P,b} = (\Delta H_f^0)_{2000,HF} = -277.57.$$

Consequently, $Q_P = 60.75 - 277.57 = -216.83$ kJ.

Generalizing what we have done if there are M different species
in both the reactants and products, the heat evolved in a constant
pressure process is

$$Q_P = \sum_{j=1}^{M} n_j [H_{T_2} - H_{298}^0 + (\Delta H_f^0)_{298}]_j$$

$$- \sum_{i=1}^{M} n_i [H_{T_1} - H_{298}^0 + (\Delta H_f^0)_{298}]_i \qquad (2.9)$$

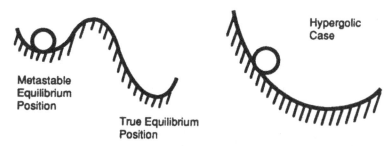

FIGURE 2.3e. Illustration of equilibria.

where the index j denotes products and the index i denotes reactants. Similarly, for a constant volume process the heat evolved is

$$Q_V = \sum_{j=1}^{M} n_j [E_{T_2} - E_{298}^0 + (\Delta E_f^0)_{298}]_j$$

$$- \sum_{i=1}^{M} n_i [E_{T_1} - E_{298}^0 + (\Delta E_f^0)_{298}]_i \qquad (2.10)$$

where the energy for any species is obtained from the enthalpy by $E_i = H_i - p/c_i$.

Before proceeding to the use of the constant pressure and constant volume heat expressions, some miscellaneous comments should be made. In the hydrogen, fluorine example, the fact is that these two reactants cannot coexist in a vessel under the stated conditions. They would immediately ignite and form the product upon coming into contact with each other. The mixture is called *hypergolic* when ignition is spontaneous. The thermodynamics are not changed, however, if we had considered the hydrogen and fluorine to be initially separated by a thin barrier which is then broken to initiate mixing and reaction. All of the energetics calculations would have been the same.

Not all fuel and oxidizer combinations are hypergolic. For example, hydrogen and oxygen can coexist in a mixture at the reference conditions of 298 K and 1 bar. In order to start reaction an ignition source is needed, such as a spark. We say that such a mixture is in a state of *metastable equilibrium*, requiring a "push" to initiate reaction. The situation is as shown in Fig. 2.3e where the ball will sit in the upper well until pushed over the hill; after such a push, the ball will go to the lower well which represents the products state. Fortunately, all thermodynamics laws apply to systems in such a state of metastable equilibrium, and their thermodynamic state may be calculated as if the mixture were in true equilibrium in the metastable state.

2.4. ADIABATIC FLAME TEMPERATURE

One of the most important calculations chemical thermodynamics allows us to make is that of the *adiabatic flame temperature*. This is the temperature achieved in a constant pressure reaction with no heat transfer. For example, reaction in a vessel which is insulated to heat transfer but operated at constant pressure would achieve the adiabatic flame temperature. Simply stated, we set $Q_p = 0$ in Eq. (2.9), and it becomes the equation describing the adiabatic process leading to the adiabatic flame temperature. Except for minor heat losses which slightly depress the flame temperature, this is the temperature achieved in jet engine and rocket combustion chambers, if the flow velocities are small (which they usually are).

Consider the constant pressure reaction between hydrogen and oxygen to form water adiabatically. From an initial state at 200 K the reaction assumed is

$$H_2 + \tfrac{1}{2}O_2 \rightarrow H_2O.$$

Equation (2.9) becomes

$$0 = (1)[H_{T_2} - H_{298}^0 + (\Delta H_f^0)_{298}]_{H_2O} - (1)[H_{200} - H_{298}^0]_{H_2}$$

$$- \tfrac{1}{2}[H_{200} - H_{298}^0]_{O_2}$$

and with numbers from the JANNAF Tables it becomes

$$0 = (H_{T_2} - H_{298}^0)_{H_2O} - 241.83 - [-2.77 - \tfrac{1}{2}(2.87)].$$

Here, T_2 is the adiabatic flame temperature. Solving for

$$(H_{T_2} - H_{298}^0)_{H_2O}$$

and going to the table for water, the result is $T_2 = 4900$ K, which is very hot!

Consider another example where hydrogen is burning with air, made up of approximately 21 moles oxygen to every 79 moles nitrogen. The reaction assumed is

$$H_2 + \frac{1}{2}\left(O_2 + \frac{.79}{.21}N_2\right) \rightarrow H_2O + \frac{1}{2}(3.76)N_2$$

and, here, the nitrogen does not enter the reaction. The nitrogen is playing the role of an *inert diluent*. The adiabatic flame temperature equation now reads

$$0 = [H_{T_2} - H^0_{298}]_{H_2O} - 241.83 + 1.88[H_{T_2} - H^0_{298}]_{N_2}$$

$$- [-2.77 - \tfrac{1}{2}(2.87)] - 1.88[H_{200} - H^0_{298}]_{N_2}$$

with two new entries from the nitrogen. T_2 is the temperature which solves this equation, but it appears in two places and an explicit solution is impossible. It would have to be solved by iteration or by trial and error, whereby several temperatures were guessed until the equation was satisfied.

Rather than taking a blind guess in the solution of the adiabatic flame equation, there is an approximate method of solution to get us close to the right answer. Viewing Table 2.3b, column 2 gives the standard state specific heat (on a molar basis) for the compound in question. In fact, these numbers are compatible with column 5 in that

$$H_{T,i} - H^0_{298,i} = \int_{298}^{T} C^0_{P,i}\, dT.$$

Now, C^0_P is a monotonically increasing function of temperature.[1] Consequently, we have

$$C^0_{P,i}(298)[T - 298] < H_{T,i} - H_{298,i} < C^0_{P,i}(T)[T - 298].$$

An upper limit estimate for the adiabatic flame temperature will therefore be attained with a lower limit estimate for C^0_P and *vice versa*. For our hydrogen–air example we may therefore estimate the T_2 by

$$(C^0_{P,H_2O} + 1.88C^0_{P,N_2})(T_2 - 298) = 243.13 \text{ kJ}.$$

From the JANNAF Tables the upper limit estimates for the specific heats at 3100 K are 56.0×10^{-3} and 37.1×10^{-3} kJ/(mol·K) for H_2O and N_2, respectively. At the lower temperature of 200 K, these numbers are 33.3×10^{-3} and 29.1×10^{-3} kJ/(mol·K), respectively. Using these estimates, an upper limit estimate for T_2 is 3049 K and the

---

[1]This is not absolutely true, but is a good approximation.

lower limit estimate is 2188 K, so we confine our accurate iteration work between these limits. By trial and error the actual answer to the adiabatic flame temperature is 2600 K, to the nearest 100 K.

Notice that the answer of 2600 K with nitrogen included is significantly lower than the 4900 K result obtained in the absence of nitrogen. The reason is that the nitrogen had to be heated up by the energy released in the hydrogen–oxygen reaction. The nitrogen behaves as a heat sink.

At this point it is useful to define what we mean by a *stoichiometric mixture*. It is a mixture of reactants that can in principle, without violating the law of stoichiometry, react to only products with the highest negative heats of formation in the chemical system being considered. Moreover, there should be no excess (leftover) fuel or oxidizer on the right side of the reaction. In the C–H–N–O system, for example, the products with the highest negative heats of formation are CO_2 and H_2O, not compounds like CO and OH. In the H–F system, HF has the highest negative heat of formation. Therefore, the following reactant mixtures are stoichiometric mixtures:

$$H_2 + \tfrac{1}{2}O_2, \quad H_2 + F_2, \quad CH_4 + 2O_2 + N_2.$$

The reason is that it is in principle possible for these mixtures to react to H_2O, HF, and the combination of CO_2 and water, respectively, with no fuel or oxidizer left over. The following reactant mixtures are not stoichiometric mixtures:

$$H_2 + O_2, \quad 2H_2 + F_2, \quad CH_4 + 10O_2 + N_2.$$

Here, even though water, carbon dioxide and hydrogen fluoride could be formed, there are either excess fuel or oxidizer.

Stoichiometric mixtures have very nearly the highest adiabatic flame temperatures possible for the chemical system under consideration. The reason is that, just as in the case of the nitrogen diluent example above, excess fuel or oxidizer acts as a diluent and cannot pass to the high negative heat of formation substances.

One other item of nomenclature will be introduced here, and that is the *equivalence ratio*. If f denotes the ratio of mass (not moles) of fuel to the mass of oxidizer in a reactive mixture, the equivalence ratio, ϕ, is

$$\phi \equiv \frac{f}{f_s}$$

where f_s is the stoichiometric mass ratio. If air is the oxidizer, convention is that the nitrogen is included in the mass of the oxidizer,

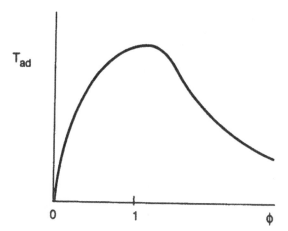

FIGURE 2.4. Adiabatic flame temperature against equivalence ratio (typical).

even though it may not react. A typical plot of adiabatic flame temperature, here given the symbol T_{ad}, versus equivalence ratio is shown in Fig. 2.4. The usual behavior of such a curve is that the maximum temperature occurs near the stoichiometric mixture ratio. The actual maximum is usually slightly to the fuel rich side of stoichiometric, for reasons that will be seen in the next section.

2.5. FREE ENERGY AND EQUILIBRIUM CONSTANT

In all the reactions considered to this point the products have been specified. They were consistent with the law of stoichiometry and were thermodynamically possible products. However, notice the extremely high temperatures calculated as adiabatic flame temperatures in the previous section. Temperature is indicative of thermal agitation of molecules, or their kinetic energy of molecular motion. If molecules collide at high kinetic energy, there is the possibility of breaking the molecules apart. Indeed this occurs, and the process is called *dissociation*. In reality, for the hydrogen–oxygen system, a mixture reacting adiabatically, or even diabatically (with heat transfer), would produce several products besides H_2O. Included would be species like H_2, H, OH, O_2, O and HO_2 (yes, the latter is possible and is known as the hydroperoxl radical). The equilibrium concentrations achieved after reaction may be small for these species, but they

CHEMICAL THERMODYNAMICS

TABLE 2.5a
Approximate flame temperatures of various stoichiometric mixtures, initial
temperature 298 K[1]

Fuel	Oxidizer	Pressure (bar)	T (K)
Carbon monoxide	Air	1	2400
Carbon monoxide	Oxygen	1	3220
Hydrogen	Air	1	2400
Hydrogen	Oxygen	1	3080
Methane	Air	1	2210
Methane	Air	20	2270
Methane	Oxygen	1	3030
Methane	Oxygen	20	3460

[1]Taken from I. Glassman, *Combustion*. Orlando: Academic Press, 1987.

exist from dissociation of the primary product, which is water. In order to make accurate calculations in thermodynamics we need to take these species into account. Fortunately, thermodynamics also tells us how to calculate which species are present, and in what abundance.

The effects of dissociation are to always lower flame temperatures, because the secondary species have higher heats of formation than the primary products. Viewing Table 2.5a, notice the actual adiabatic flame temperature of a stoichiometric mixture of hydrogen and oxygen. It is 3080 K, not the 4900 K calculated in the previous section.

The second law of thermodynamics, also true for a chemically reacting system, introduces the entropy which behaves as

$$T\, d\hat{S} \geq d\hat{E} + p\, dV \tag{2.11}$$

for a fixed mass system, with the equality holding at equilibrium (no tendency for spontaneous chemical change). Thus, the entropy tends to increase with chemical change if $d\hat{E}$ and $d\hat{V}$ are zero (a constant energy and volume process) and will achieve a maximum at the equilibrium condition in such a process. Since the energy, $\hat{E}$, is a somewhat abstract variable, we wish to consider some more physical variables to describe the process at chemical equilibrium. Consider the grouping of variables $E + pV - TS$, and consider a differential process which holds pressure and temperature constant. That is, using inequality (2.11),

$$d(\hat{E} + pV - T\hat{S}) = d(\hat{H} - T\hat{S}) = d\hat{E} + p\, dV - T\, d\hat{S} \leq 0.$$

We define the variable $\hat{H} - T\hat{S}$ as a new variable, $\hat{G}$, called the free energy (actually, the Gibbs' free energy) which compactly states the above inequality as

$$d\hat{G} \leq 0, \quad p \text{ and } T \text{ constant}$$

and states that the free energy must achieve a minimum at the point of chemical equilibrium.

Just as with $\hat{H}$ and $\hat{E}$, the free energy and entropy for a fixed mass system may be computed as a sum of the energies and entropies of the constituent species. That is,

$$\hat{G} = \sum_{i=1}^{M} n_i G_i, \qquad \hat{S} = \sum_{i=1}^{M} n_i S_i.$$

The entropies for perfect gases are computed from

$$S_i = S_i^0(T) - R \ln\left(\frac{p_i}{p_{\text{ref}}}\right) \tag{2.12}$$

with the standard state entropy given by

$$S_i^0 \equiv \int_0^T \left(\frac{C_{P,i}^0}{T}\right) dT.$$

Notice in column 3 of Table 2.3b the latter integral is given. Now take p_{ref} as 1 bar and *always work in bars from now on* or the units will not be right. This move simplifies Eq. (2.12) and is merely a convention, but it is used in the JANNAF Tables as well as in many other sources of thermochemical data.

Now consider the stoichiometric reaction used before.

$$H_2 + \tfrac{1}{2}O_2 \rightarrow H_2O.$$

In reality, this reaction will form more products than just H_2O. Assume, for the moment, that the following reaction is more realistic:

$$H_2 + \tfrac{1}{2}O_2 \rightarrow n_{H_2O}H_2O + n_{H_2}H_2 + n_{O_2}O_2.$$

The law of stoichiometry is insufficient to determine the n_i, but an atomic balance yields

$$n_{H_2O} = 1 - n_{H_2}, \qquad n_{O_2} = \tfrac{1}{2}n_{H_2}.$$

We now inquire as to what n_{H_2} (and consequently n_{H_2O} and n_{O_2}) is if the product mixture is in equilibrium. It is important to recognize that we consider only the product side of the reaction of interest. All the consideration of the reactants is now dropped. What happens is that the n_i will adjust themselves to meet the condition of equilibrium satisfying the minimization of $\hat{G}$. The adjustment occurs by forward and reverse reactions on the product side. In the example case the only stoichiometrically correct reaction which can take place between the assumed product species is

$$H_2 + \tfrac{1}{2}O_2 \rightleftarrows H_2O.$$

Showing the reaction going both ways means that a balance will take place; at equilibrium the forward (right) rate of production of water will be balanced by an equal rate of destruction of water to form hydrogen and oxygen so that the composition remains stable.

We now form the $\hat{G}$ for the vessel containing all three species. It is, using Eqs. (2.11) and (2.12),

$$\hat{G} = \sum_{i=1}^{M} n_i H_{T,i} - T\sum_{i=1}^{M} n_i S_i = \sum_{i=1}^{M} n_i[(H_{T,i} - TS_i^0 + RT\ln p_i)]$$

$$= \sum_{i=1}^{M} n_i[G_i^0 + RT\ln p_i] \tag{2.13}$$

where the definition of the standard state free energy (G at one bar, but at any T of interest) is made that

$$G_i^0(T) \equiv H_{T,i}^0 - TS_i^0.$$

Recall that perfect gases are presumed so that $H_{T,i} = H_{T,i}^0$ in the above. Now the only variable in Eq. (2.13) is n_{H_2}, if T and p are held constant; the p_i depend on n_i through Eqs. (2.3) and (1.1), with $c_i = n_i/V$. Consequently, a differential change in $\hat{G}$ can only take place through a differential change in n_{H_2}. Computing this derivative with constant p and T and setting it equal to zero at equilibrium, we have

$$\frac{d\hat{G}}{dn_{H_2}} = (1)G_{H_2}^0 + \left(\frac{1}{2}\right)G_{O_2}^0 - (1)G_{H_2O}^0 + RT\frac{d}{dn_{H_2}}\sum_{i=1}^{M} n_i\ln p_i = 0.$$

Note that the first three terms are the *negative* of the change of G^0 for the formation of water from the elements hydrogen and oxygen. It is called the standard state free energy of formation and is given the symbol $\Delta G^0_{f,H_2O}$. It is listed in column 7 of the JANNAF Tables for any compound or atom. It is, of course, zero for a reference element. Continuing, note that

$$\frac{d}{dn_{H_2}} \sum_{i=1}^{M} n_i \ln p_i = (1)\ln p_{H_2} + \left(\frac{1}{2}\right) \ln p_{O_2}$$

$$- (1)\ln p_{H_2O} + \sum_{i=1}^{M} n_i \frac{d}{dn_{H_2}} \ln p_i$$

and

$$\sum_{i=1}^{M} n_i \frac{d}{dn_{H_2}} \ln p_i = \sum_{i=1}^{M} \frac{n_i}{p_i} \frac{dp_i}{dn_{H_2}} = \frac{V}{RT} \sum_{i=1}^{M} \frac{d}{dn_{H_2}} p_i$$

$$= \frac{V}{RT} \frac{d}{dn_{H_2}} \sum_{i=1}^{M} p_i = \frac{V}{RT} \frac{dp}{dn_{H_2}} = 0$$

since p is constant. When we put this all together,

$$\frac{d\hat{G}}{dn_{H_2}} = 0 = -\Delta G^0_{f,H_2O} + RT \ln \left(\frac{p_{H_2} p_{O_2}^{1/2}}{p_{H_2O}} \right)$$

or

$$\frac{p_{H_2O}}{p_{H_2} p_{O_2}^{1/2}} = \exp\left[-\frac{\Delta G^0_{f,H_2O}}{RT} \right] = f(T \text{ alone}) \equiv K_p(T), \quad (2.14)$$

where the function K_p depends upon temperature alone. It is called the *equilibrium constant* based upon partial pressures. It is sometimes confusing that partial pressures in Eq. (2.14) are involved, but K_p is only temperature dependent. It is the particular grouping of the partial pressures in Eq. (2.14) that is only dependent upon temperature, not pressure. The exponents on the partial pressures are stoichiometric coefficients of the reaction, as initially written, with the exponents in the numerator those of the products and in the denominator those of the reactants.

The reader may verify that if any arbitrary reaction is written, such as

$$aA + bB + \cdots \rightleftharpoons rR + sS + \cdots$$

and if the above analysis is carried out using the perfect gas assumption, the equilibrium will always be expressed as

$$\frac{p_R^r p_S^s \cdots}{p_A^a p_B^b \cdots} = K_p(T) = \exp\left[\frac{-\Delta G^0}{RT}\right]. \qquad (2.15)$$

Remember, however, that pressures must be expressed in the units of bars; otherwise, the starting point for the entropy is wrong.

Column 8 of the JANNAF Tables contains the logarithm to the base 10 of K_p for the formation reaction of any compound. It is fortunate that the equilibrium constant for any arbitrary reaction may be expressed in terms of the formation K_{pf} for all entering substances. For example, consider the equilibrium of the reaction

$$H_2O \rightleftharpoons 2H + O.$$

Manipulating,

$$K_p = \left(\frac{p_H^2 p_O}{p_{H_2O}}\right) = \left(\frac{p_H}{p_{H_2}^{1/2}}\right)^2 \left(\frac{p_O}{p_{O_2}^{1/2}}\right) \left(\frac{p_{H_2} p_{O_2}^{1/2}}{p_{H_2O}}\right)$$

$$= \frac{(K_{pf,H})^2 (K_{pf,O})}{(K_{pf,H_2O})}$$

Consequently, the JANNAF listing of the formation equilibrium constants is all that is necessary for arbitrary reactions. Notice that K_{pf} for reference elements is unity (or $\log_{10} K_{pf} = 0$) because the formation reaction for the elements is merely an identity (i.e. reference element → reference element).

Example 2.2.
If it is assumed that the following reaction takes place and that the final pressure is 1 bar with a temperature of 3500 K, what is the distribution of products?

$$H_2 + F_2 \rightarrow n_{HF}HF + n_H H + n_F F.$$

Solution.

From an atomic balance

$$2 = n_{HF} + n_H, \qquad 2 = n_{HF} + n_F$$

which yields

$$n_H = n_F \qquad \text{and} \qquad n_{HF} = 2 - n_H.$$

The formation reactions for the substances formed and their corresponding equilibrium constants are

$$\tfrac{1}{2}H_2 + \tfrac{1}{2}F_2 \rightarrow HF, \qquad K_{pf,HF} = \left(\frac{p_{HF}}{p_{H_2}^{1/2} p_{F_2}^{1/2}} \right) = 10^{4.221},$$

$$\tfrac{1}{2}H_2 \rightarrow H, \qquad K_{pf,H} = \left(\frac{p_H}{p_{H_2}^{1/2}} \right) = 10^{-.228},$$

$$\tfrac{1}{2}F_2 \rightarrow F, \qquad K_{pf,F} = \left(\frac{p_F}{p_{F_2}^{1/2}} \right) = 10^{2.121}.$$

The total number of moles in the products is

$$n = n_{HF} + n_H + n_F = 2 + n_H.$$

For the reaction $H + F \rightleftarrows HF$ the equilibrium is stated by

$$K_p = \left(\frac{p_{HF}}{p_H p_F} \right) = \frac{K_{pf,HF}}{K_{pf,H} K_{pf,F}} = \left(\frac{n}{p} \right) \frac{n_{HF}}{n_H n_F} = 10^{(4.2-2.1+.231)},$$

$$\frac{(2 + n_H)(2 - n_H)}{n_H^2(1)} = 10^{2.331} = 214,$$

$$n_H = 0.0187 = n_F \qquad n_{HF} = 1.9813,$$

or, approximately,

$$X_H = 0.0093 = X_F, \qquad X_{HF} = 0.9814.$$

Example 2.3.

Ammonia (NH_3) is an important industrial chemical, which has wide use. It is produced in the Haber–Bosch process at high temperature (550 K) by its formation reaction in the presence of a catalyst. Under these conditions, what is the yield of NH_3 (its mole fraction), given that the free energy of formation is 8.4 kJ/mol?

Solution.

The formation reaction is

$$\tfrac{1}{2}N_2 + \tfrac{3}{2}H_2 \rightarrow NH_3.$$

The equilibrium constant of formation is

$$K_{pf,NH_3} = \frac{p_{NH_3}}{(p_{N_2}^{1/2} \cdot p_{H_2}^{3/2})} = \exp\left[\frac{-8.4}{(8.314 \times 10^{-3} \cdot 550)}\right] = 0.160.$$

By an atomic balance, assuming only N_2, H_2 and NH_3 are present

$$2n_{H_2} + 3n_{NH_3} = 3, \qquad 2n_{N_2} + n_{NH_3} = 1,$$

$$n_{N_2} = (1 - n_{NH_3})/2, \qquad n_{H_2} = \tfrac{3}{2}(1 - n_{NH_3}),$$

and

$$n = 2 - n_{NH_3}$$

gives partial pressures as

$$p_{NH_3} = p n_{NH_3}/n,$$

$$p_{N_2} = p(1 - n_{NH_3})/(2n),$$

$$p_{H_2} = 3p(1 - n_{NH_3})/(2n)$$

or

$$0.160(300) = n_{NH_3}(2 - n_{NH_3})\{[\tfrac{1}{2}(1 - n_{NH_3})]^{1/2}[\tfrac{3}{2}(1 - n_{NH_3})]^{3/2}\}$$

yielding by trial and error

$$n_{NH_3} = 0.875, \qquad X_{NH_3} = 0.778.$$

Notice in this example the role of the high pressure. It is the only way to get the high yield. At this pressure the gases are actually behaving imperfectly, but the numbers calculated here are of the correct order of magnitude.

The effect of pressure on the way the equilibrium composition comes out is obtained by setting $p_i = X_i p$ in the general reaction equation, Eq. (2.15). That is,

$$\frac{X_R^r X_S^s}{X_A^a X_B^b} = K_p(T)p^{(a+b+\cdots-r-s-\cdots)} = K_p p^z.$$

Recall that the composition is specified by the X_i. The pressure effect is measured by $z = (a + b + \cdots) - (r + s + \cdots)$. If this quantity, z, is zero, the mole fractions will be unique regardless of pressure. If z is positive the X_i in the numerator must fall if the pressure rises, and *vice versa*. That is, if z is positive, the species on the left side of the reaction equation are favored (the equilibrium reaction shifts to the left). The origin of this effect lies in whether or not there are more moles of gases on the right side of the reaction equation as compared with those on the left. If there are then $z < 0$. If there are an equal number of moles on both sides of the reaction equation then $z = 0$. Therefore, the rule is that under increasing pressure the equilibrium shifts so that the total number of moles (molecules) decreases.

The effect of temperature is a little more complex to reason, but, approximately, atoms or molecules with a positive heat of formation are favored as the temperature rises, and *vice versa*. As an example, in a mixture of H_2 and H at fixed pressure the production of more H is favored as the temperature rises. This occurs because at higher temperature there is higher random thermal motion of the gas molecules and the H_2 dissociates more.

It is imperative to realize in calculation of chemical equilibria that all conceivable reactions in the system must be in equilibrium. For this reason, as in the examples above, we have complete liberty in choosing the reactions to consider in making equilibrium calculations. The ones we choose are taken only for computational convenience. We are now in a position to return to calculation of adiabatic flame temperature through use of Eq. (2.9), with $Q_p = 0$. Now we know how to calculate the n_j, which in general depend not only upon the pressure of interest, but also the adiabatic flame temperature itself. That is, $n_j = n_j(p, T_2)$, so that at a fixed pressure Eq. (2.9) is a equation for T_2, *albeit* a complicated one. The following procedure is a "cookbook" method for making adiabatic flame temperature calculations:

1. Write down the initial reaction equation with the reactants going to the products that we wish to consider. Some experience is required here in choosing the dominant products, but we can check whether proper assumptions were made after the initial calculation is finished, as seen later.

2. From an atomic balance, reduce the number of unknown n_j. If there are M values of n_j and R different atomic types in the reactants, there are now $M-R$ unknowns for the stoichiometric coefficients on the product side.

3. Now throw away consideration of the initial reactants, and concentrate on the equilibrium among the product species. Write down the equilibrium laws (from the formation reactions) for all compounds occurring in the products which are not reference state elements. Only do this for compounds for which the n_j are still unknown from step 2. It will now often be seen that partial pressures appear for compounds which are not considered to be products in the problem. For example, we might consider that H is a product but, for some reason, have excluded H_2. The formation reaction for H involves H_2, however.

4. It is always possible, by algebraic manipulation, to eliminate the partial pressures of unwanted species by division and multiplication of the formation equilibrium equations. The equilibrium equations may always be reduced to $M-R$ equations containing the unknowns.

5. The adiabatic flame temperature is unknown. Guess one. Solve the equilibrium equations and obtain the n_j. This algebraic process is always highly nonlinear.

6. With the known n_j and guessed T_2 evaluate the adiabatic flame equation. In all probability it will not be satisfied.

7. Guess another temperature and repeat steps 5 and 6. Continue this process until the adiabatic flame temperature equation is satisfied. This is the answer, if we are satisfied with the products chosen.

8. An after-the-fact check may now be made on excluded products. For example, if we have excluded H_2 but included H, we may now write down the formation reaction for H and with the computed T_2 compute p_{H_2}. If it is negligible compared with p_H we are finished. Otherwise H_2 would have to be included in the overall computation.

The reader should go through this procedure on Example 2.2 above. In general, for realistic combustion problems, the calculation is too laborious to do by hand because of the large number of product species which are important. Fortunately, there are several computer programs now which can rapidly compute very complex equilibria problems and yield adiabatic flame temperatures, as well as make other propulsion and explosion calculations. Some of these programs are available for rather modest memory personal computers.[1] The actual method of computation may proceed through a minimization of the free energy or through equilibrium constants, but the data base used is always the JANNAF Tables.

In making hand calculations there is one convention which should be pointed out, and that has to do with carbon as well as any other reference state element which is in a condensed phase. The formation reaction for CO_2, for example, is

$$C(s) + O_2(g) \rightarrow CO_2(g)$$

and it is tempting to write the equilibrium constant for this reaction as

$$K_{pf,CO_2} = \frac{p_{CO_2}}{(p_C p_{O_2})}.$$

But what is p_C? Carbon is in the solid phase. The answer to this dilemma is that every solid has a *vapor pressure* of its own vapor with which it can be in equilibrium. That is, the *sublimation* equilibrium reaction may be written

$$C(s) \rightleftarrows C(g).$$

The p_C in the equilibrium relation above is, in fact, this vapor pressure. But the convention in the JANNAF Tables is to absorb this number into K_{pf}, yielding the tabulated quantity as

$$K' = K_{pf,CO_2} p_c = p_c \exp\left[\frac{-(\Delta G_f^0)_{CO_2}}{RT}\right].$$

---

[1] The references for some of these programs are: (a) S. Gordon and B. J. McBride, NASA SP-273, 1971; (b) D. R. Cruise, NWC TP 6037; (c) M. Cowperthwaite and W. H. Zwisler, SRI Publication No. Z106, Vol. IV, 1974. NWC stands for Naval Weapons Center and SRI stands for Stanford Research Institute. The computer programs for (b) and (c) are NEWPEP and TIGER. The program for (a) is called TRAN76. These programs may be obtained from the cited organizations.

2.6. PROBLEMS

1. A vessel contains a mixture of perfect gases at $p = 2$ bar and $T = 400$ K. The mixture consists of 1 lb H_2, 10 mol F_2 and enough water vapor to give a partial pressure of 0.5×10^5 Pa. Calculate (a) the mass fractions of all species, (b) their mole fractions, (c) their concentrations, (d) their partial pressures and (e) the density of the mixture.
(Answer: $Y_{H_2} = 0.20$, $Y_{F_2} = 0.17$, $Y_{H_2O} = 0.63$)

2. Verify by numerical integration from 1000 to 1100 K in Table 2.3a that the change in column 5 is equal to the integral of column 2 over the temperature. Also verify that if column 2 is divided by T and integrated it is equal to the change in column 3.

3. Calculate $(\Delta H_f^0)_{1000,O}$, for the O atom, using the heat of formation at 298 K and the appropriate sensible enthalpy data for O and O_2.

4. Calculate the adiabatic flame temperature for the reaction

$$H_2 + O_2 \rightarrow n_{H_2O}H_2O + n_{O_2}O_2 + n_O O$$

with the reactants initially at 298 K. Do this for $p = 1$ and 10 bar. Using the estimated temperature make an appropriate calculation of the amount of H_2 present to see whether or not the neglect of the presence of this substance is justified.
(Answer: $T_2 = 3200$ K)

5. If the reference temperature of absolute zero is chosen instead of 298 K, what is the value of the energy of formation for perfect gases in terms of the enthalpy of formation?

6. Methane is burned with air at an equivalence ratio of 1.2. The process is a steady flow reaction process at a constant pressure of 1 bar. The initial temperature is 298 K and the total flow rate of methane and air is 1 kg/s. If the products are assumed to be CO, H_2O, CO_2, and N_2, and the process is adiabatic, what is the final temperature?
(Answer: $T_2 = 1950$ K)

CHEMICAL
KINETICS

3.1. INTRODUCTION

All of the chemical reactions shown as examples in previous chapters almost never really take place as written. The true path of product creation is usually a complex one involving many steps at the molecular or atomic level. From the standpoint of thermodynamics, it did not matter that the true reaction mechanism was in error, because only the initial and final states were of interest. But often we need the true reaction path to study flame phenomena. We also need information on the rate in proceeding from an initial state to a final state. Such rates of chemical reactions are furnished by the science of *chemical kinetics*.

The determination of actual reaction paths involves very complex inference from usually incomplete data and theoretical argument. However, at this time, there are many reaction schemes that are quite well known; there are also many others which are known quite incompletely. From an engineering viewpoint, complete knowledge of a reaction scheme is often not necessary, and useful approximations can be made. Indeed, in this text we shall use mostly approximate approaches, since they can yield valuable insight into flame behavior.

3.2. REACTION RATE

We take it as axiomatic that atoms or molecules cannot react unless they collide.[1] We also take it as intuitive that the rate of atomic or molecular collisions must be proportional to the number densities (number per unit volume) of the colliding species. For example, in the reaction

$$O + H \rightarrow OH$$

the rate of reaction for the reactants must be proportional to $c_H c_O$. We read the reaction above as "one atom of oxygen collides with one atom of hydrogen to form one molecule of OH." We do not use moles or have fractional stoichiometric coefficients when dealing with actual chemical reaction mechanisms. We define the reaction rate, R, through a coefficient, k, called the *specific reaction rate constant*, by

$$R = k c_H c_O$$

for the above reaction. For a general reaction, we multiply k by the product of the concentrations of the reacting species. For example, for

$$H + H + H \rightarrow H_2 + H,$$

$R = k c_H^3$. The reaction rate tells how often a reaction occurs and is used to tell the time rate of concentration change of the reacting species. For example, in the first reaction above, the units of k are

---

[1]This is not true, in general. Another mechanism of reaction can be the absorption of electromagnetic radiation. However, we shall not be concerned with this mechanism in this book.

given by the following equality:

$$\frac{dc_O}{dt} = -(1)kc_O c_H = -R$$

with the minus sign being chosen because O is being destroyed. In the second example above, because H is both being destroyed by the reaction and created on the right side, the concentration change of H is

$$\frac{dc_H}{dt} = -3R + R = -2R = -2kc_H^3.$$

Alternatively, for the H_2,

$$\frac{dc_{H_2}}{dt} = +R = kc_H^3$$

with the plus sign because H_2 is being created. The rate of the reaction is given by the product of k and the reactant concentrations whereas the numerical coefficient tells the net molecular number change of a given species in each reaction.

The above two cases are examples of *two-body* and *three-body* reactions, respectively, indicating a collision between either two molecules or three molecules (or atoms, here) must take place before a reaction can occur. Notice, in order for the units to be correct, the units of k depend upon the number of bodies colliding. Two- and three-body reactions are the most common ones encountered. The probability that four or more bodies could get together is simply too low, especially in perfect gas systems.

Experimentally, if a true reaction is being studied, the behavior of Fig. 3.2a is found. The straight line on such a plot is given by the analytical form

$$k = Ae^{-B/T} \tag{3.1}$$

with A being called the *pre-exponential factor*. Because there is theoretical support for such a law, $B = E/R$, with E being called the *activation energy*. E may be thought of as the minimum energy required in a collision for the reaction to, in fact, occur. The pre-exponential factor can have a weak temperature dependence, since collision rates are so dependent.[2] The law expressed by Eq. (3.1) with $B = E/R$ is called the *Arrhenius law*.

---

[2]There are other reasons for a slight dependence of A upon T, and E may have a slight T dependence. However, for purposes here, such dependencies will not be explored.

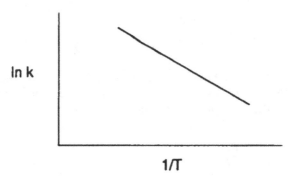

FIGURE 3.2a. Experimental behavior of the reaction rate constant when the Arrhenius law is obeyed.

The sum of the exponents on the concentrations in the reaction rate is called the *molecularity* of the reaction, and it indicates the number of particles entering the reaction. For two-body reactions the molecularity is two. From Eq. (2.3)

$$c_i = \frac{p_i}{RT} = \frac{X_i p}{RT}$$

so that, for fixed composition and temperature, c_i is proportional to pressure. Whenever we discuss the pressure dependence of reaction rates, from here on, it is assumed that T and the X_i are fixed. Consequently, two-body reactions have a rate proportional to p^2, and three-body reactions have a rate proportional to p^3. As pressure increases, therefore, three-body reactions increase in rate faster than do the rates of two-body reactions.

One other type of reaction occurs with highly reactive species in enclosures. Some species may be absorbed by walls. In order for this to occur, the species must have diffused to the proximity of the wall and then collided with it. The difficulty of migration to a wall by random molecular migration intuitively depends on how crowded the molecules are. That is, if the subject molecule bumps into a lot of other molecules it would appear to take longer to get to a wall than if the motion were unimpeded. The situation is as illustrated in Fig. 3.2b. Therefore, the time to get near a wall is proportional to ρ, which, in turn, is proportional to p. Alteratively, the rate at which molecules appear near the wall is proportional to $1/p$. Moreover, once near a wall, the collision rate with a wall should be proportional to p, because p is proportional to number density. Both of these arguments

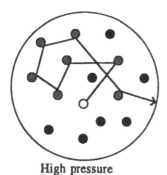
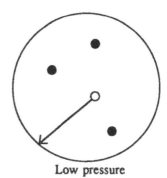

High pressure   Low pressure

FIGURE 3.2b. Illustration of difficulty in reaching the wall for a molecule in a high and low pressure (density) case.

support the contention that wall reaction rates are dependent upon pressure to no larger than the first power. The consequence is that three reaction types (wall, two-body, and three-body) depend upon pressure in different ways, and this will be useful later in reasoning how explosion phenomena depend upon pressure.

The reactions written above indicated only left to right reaction. This is called the forward reaction. But the reverse can also occur. We denote the rate constants as k_f and k_b for the forward and backward (reverse) reactions, respectively. For example,

$$H + H + H \underset{k_b}{\overset{k_f}{\rightleftarrows}} H_2 + H$$

with

$$\frac{dc_H}{dt} = -2k_f c_H^3 + 2k_b c_{H_2} c_H.$$

We have already noted that $k_f = k_f$ (T alone). The same is true for k_b. Note the special case of equilibrium where $dc_H/dt = 0$. Then it must be true that, using Eqs. (2.3) and (2.15),

$$\frac{c_{H_2}}{c_H^2} = \frac{k_f(T)}{k_b(T)} = RT\frac{p_{H_2}}{p_H^2} = RTK_p(T) = \frac{RT}{K_{pf,H}^2}.$$

That is, since the functions of temperature are related through the equilibrium constant and functions cannot change even when out of equilibrium (net reaction occurring), only one of the reaction rate constants is independent. If either the forward or the reverse reaction rate constant is measured (or theoretically deduced) the other is determined through thermodynamics.

Example 3.1.

Consider the hypothetical reaction

$$O + CO_2 + H_2O \underset{k_b}{\overset{k_f}{\rightleftharpoons}} OH + CO + O + O + H.$$

Write the reaction rate for the forward and reverse reactions. Calculate the net production rate of the oxygen atom and water in moles per unit volume per unit time.

Solution.

$$\mathbf{R}_f = k_f c_O c_{CO_2} c_{H_2O},$$

$$\mathbf{R}_b = k_b c_{OH} c_{CO} c_O^2 c_H,$$

$$\frac{dc_O}{dt} = (2-1)\mathbf{R}_f + (1-2)\mathbf{R}_b = \mathbf{R}_f - \mathbf{R}_b,$$

$$\frac{dc_{H_2O}}{dt} = -(1)\mathbf{R}_f + (1)\mathbf{R}_b = \mathbf{R}_b - \mathbf{R}_f.$$

3.3. RADICALS

Compounds which are highly stable at normal temperatures usually emulate the noble gases (helium, neon, argon, etc.) in the configuration of their electrons. That is, these highly inert gases have outer electron shells empirically "filled" by 2, 8 and 8 electrons in the case of He, Ne and A, respectively. Compounds which attempt to mimic this situation are usually quite stable and are hard to break apart. An example is diatomic H_2, which looks like He in the sense that each hydrogen atom shares its electron with the other atom so that there are two electrons in the configuration. Similarly, H_2O looks like Ne where there are a total of 8 electrons being shared, 6 from the outer shell in oxygen and 2 from both of the hydrogen atoms.

Any atom or compound with an unfilled outer electron configuration is highly reactive and is called a *radical*. Examples are the hydroxyl radical, OH, and the oxygen and hydrogen atoms, O and H. While an equilibrium calculation at usual temperatures shows that

TABLE 3.3a

Some reaction rate parameters[1] for the hydrogen–oxygen system, $k = A'T^n e^{-E/RT}$. The units are consistent with production of a dc_i/dt in $(mol)(cm^{-3})(sec^{-1})$ with T in Kelvin and E in kcal/mol

Reactions	$\log_{10} A'$	n	E
$H + O_2 \rightarrow O + OH$	16.71	−0.816	16.51
$O + H_2 \rightarrow H + OH$	10.26	1.0	8.90
$H_2 + OH \rightarrow H_2O + H$	13.34	0.0	5.15
$O + H_2O \rightarrow OH + OH$	13.83	0.0	18.36
$HO_2 + OH \rightarrow H_2O + O$	13.70	0.0	1.00
$H + O_2 + M \rightarrow HO_2 + M$	15.18	0.0	−1.00
$H + OH + M \rightarrow H_2O + M$	23.19	−2.0	0.00

[1]From C. K. Westbrook and F. L. Dryer, *Prog. Energy and Combust. Sci.* **10** (1984), p. 1. Note that the kilocalorie is the energy unit here.

radicals are only present in small amounts, they play a central role in chemical kinetics reaction schemes. Table 3.3a lists a few reactions which take place in the hydrogen–oxygen system. Seen there are the hydroperoxyl radical, HO_2, OH and the atoms H and O. During the conversion from reactants to products in combustion reactions, radicals are crucial to the reaction path.

Radicals usually have positive heats of formation, which means that energy is required for their formation from the standard state elements. They soak up collisional energy in their formation and wish to give up this energy in attacking other species in the path toward reaction completion. Hungry for electrons, radicals are efficient intermediates in reaction schemes for taking the reaction path forward. It should be pointed out that the above reasoning concerning the electron configuration is not always straightforward. For example, a complex electron sharing occurs with CO, and it is not a radical, although it would appear to be one.

Radicals are extremely important in the kinetic schemes relevant to atmospheric pollution. Examples of reactions producing nitric oxide (NO), a deadly poison present as an air pollutant, are

$$O + N_2 \rightarrow NO + N, \qquad N + O_2 \rightarrow NO + O.$$

These two reactions comprise what is called the thermal mechanism for formation of NO. There are several other mechanisms, all involving radicals.

3.4. EXPLOSION EXAMPLE—THE HYDROGEN–OXYGEN EXPLOSION

Under conditions of low temperature and pressure it has already been mentioned in Chapter 2 that gaseous hydrogen and oxygen are in a metastable equilibrium and will not react. However, if the p or T are raised sufficiently high a violent exothermic reaction will be initiated as the product, water, is formed. The reaction is also fast, compared with time scales that humans usually consider as ordinary. While it is difficult to define precisely, such a fast, exothermic reaction is called an *explosion*. In the case of hydrogen and oxygen, this explosion can be spontaneous, without requiring any ignition source, such as a spark. Qualitative reasoning with a rather simple chemical kinetics scheme can give insight into the H_2–O_2 explosion and its dependence upon pressure and temperature conditions.

It might seem reasonable that the reaction scheme

$$H_2 + H_2 + O_2 \rightarrow H_2O + H_2O$$

would explain the formation of water. It never occurs, however. First of all, it would require a three body collision that has the three bodies perfectly aligned in the proper orientation to form the two water molecules. Second, it would require that all diatomic bonds be broken. Third, since this is a highly exothermic reaction, where would the energy go? It cannot go anywhere, and it would merely go to breaking up the newly formed bonds. A much more complex scheme is involved in the oxidation of the hydrogen.

Figure 3.4a shows the pressure and temperature conditions under which a spontaneous explosion will occur.[3] This is shown for a particular containment vessel, and the explosion conditions do depend upon the wall material and vessel size. The behavior is reasonably complicated; at any T, above a certain minimum, there are three pressure limits separating an explosion region from a non-explosion region.

The minimum number of individual reactions which can explain this behavior is given in the following list:

1. $M + H_2 \rightarrow M + H + H$
2. $H + O_2 \rightarrow OH + O$

---

[3]From B. Lewis and G. von Elbe, *Combustion, Flames and Explosions of Gases*, 2nd ed. New York: Academic Press, 1961, p. 24.

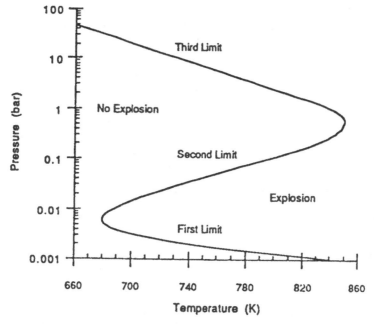

FIGURE 3.4a. Explosion limits of a stoichiometric hydrogen–oxygen mixture in a spherical KCl-coated vessel of 7.4 cm diameter.

3. $O + H_2 \rightarrow OH + H$
4. $OH + H_2 \rightarrow H_2O + H$
5. $H \rightarrow$ wall
6. $OH \rightarrow$ wall
7. $H + O_2 + M \rightarrow HO_2 + M$
8. $HO_2 + H_2 \rightarrow H_2O + OH$
9. $HO_2 \rightarrow$ wall

The first four reactions and reaction (8) are bimolecular reactions and their rates will scale with pressure squared. Reaction (7) is trimolecular and will increase its rate in proportion to pressure cubed. The wall reactions scale with pressure to no larger than the first power. Now note that in order for the water to be formed H and OH must be created by reactions (1)–(4) or HO_2 must be formed in reaction (7). At very low pressure any radical formation will be met with wall destruction, and no explosions may occur. Raising pressure from this condition raises the rates of reactions (1)–(4) relative to the wall destruction reactions and eventually the lower explosion limit is reached.

It is possible to start with a higher pressure yet in the non-explosion region (above the second limit, but below the third limit) because of the emergence of reaction (7), which competes with reaction (2) for the H atom. Under these conditions, however, when HO_2 is formed it becomes destroyed by the wall. Lowering the pressure, lowers the rate of reaction (7) with respect to reactions (1)–(4) and the second limit will be encountered. On the other hand, raising the pressure from this point raises the rate of reaction (7) relative to the wall destruction reaction (9). Ultimately, the third limit is reached.

It is remarkable that such a simple reaction scheme can explain such complex behavior. Most fuel and oxidizer combinations have much more complicated chemical kinetics schemes. Nevertheless, they all have explosion limits which vary in their characteristics, as compared with the limits in the H_2–O_2 system. Other books cover the oxidation phenomena of fuels in much more detail than will be done here.[4]

3.5. GLOBAL KINETICS

A detailed listing of reactions involving hydrocarbons and their rates may be found elsewhere.[5] For engineering purposes, however, it is often possible to bypass the details of chemical kinetics and still retain some information on scaling laws for many flame types. A technique called global kinetics may often be used, whereby an approximate chemical rate expression may be used. For example, consider the stoichiometric reaction

$$CH_4 + 2O_2 \rightarrow CO_2 + 2H_2O$$

which, although it never occurs, is thermodynamically possible. In global kinetics we presume that the overall product formation rate can be approximated by a rate expression of the form

$$\frac{dc_{CO_2}}{dt} = c_{CH_4}^p c_{O_2}^q A e^{-E/RT}.$$

We then attempt to empirically determine the constants A, p, q and E for a best fit of such a law to the data. We still call A and E the

---

[4]For example, see I. Glassman, *Combustion*, 2nd ed. Orlando: Academic Press, 1987.
[5]See C. K. Westbrook and F. L. Dryer, *Prog. Energy and Combusti. Sci*, 10 (1984), p. 1.

pre-exponential factor and activation energy, respectively, but know that they are not the factors introduced earlier, because a true reaction mechanism is not being considered. The constants p and q, also empirically determined, need not be integers, since a real molecular encounter is not being considered. It is intuitive that p and q should be positive because it seems reasonable that the greater the concentrations of the reactants the faster should be the conversion rate.

We define $n = p + q$ as the apparent *order* of the reaction, and it gives the apparent number of molecules entering the reaction. However, again, because of the approximation n need not be integer as it would be if a real reaction step were being considered.

We are at liberty to get as complex as desired in global kinetics.[6] If we find that a single reaction step does not fit the rate data very well, then a multistep procedure might be adopted. For the above reaction we could break it up into two steps, for example, as

$$CH_4 + 2O_2 \xrightarrow{k_1} CO + 2H_2O + \tfrac{1}{2}O_2, \qquad \frac{dc_{H_2O}}{dt} = k_1 c_{CH_4}^{p_1} c_{O_2}^{q_1},$$

$$CO + \tfrac{1}{2}O_2 \xrightarrow{k_3} CO_2, \qquad \frac{dc_{CO_2}}{dt} = k_2 c_{CO}^{p_2} c_{O_2}^{q_2}.$$

Of course, increasing the number of reactions increases the analytical complexity.

Example 3.2.
Effective global schemes may sometimes be concocted by making partial equilibrium assumptions. For the reaction scheme

$$H_2 + O_2 \underset{k_2}{\overset{k_1}{\rightleftarrows}} 2OH,$$

$$H_2 + OH \xrightarrow{k_3} H_2O + H$$

assume the first reaction to be in equilibrium and calculate the water production rate from the third.

---

[6]See for example, F. L. Dryer, *Fossil Fuel Combustion: A Source Book* (Bartok and Sarofim, eds.). New York: Wiley, 1991, p. 121.

Solution.

$$K_{p_{12}} = P_{OH}^2/(P_{H_2}P_{O_2}) = c_{OH}^2/(c_{H_2}c_{O_2}),$$

$$\frac{dc_{H_2O}}{dt} = k_3 c_{H_2} c_{OH} = k_3 K_{p_{12}}^{1/2} c_{H_2}^{3/2} c_{O_2}^{1/2}$$

and $k_3 K_{p_{12}}^{1/2} = k'$ becomes an effective global raction rate constant.

3.6. PROBLEMS

1. For the reactions in Table 3.3a, write the expressions for the time rate of concentration change for the first substance on the right side of the reaction. The result should be in terms of the reactant concentrations and the specific reaction rate constants.

2. For an elementary reaction you have found the activation energy to be 80 kJ/mol and the pre-exponential factor to be independent of temperature. If the temperature is doubled from 1000 K to 2000 K by what factor does the reaction rate increase?

3. One mechanism for the oxidation of CO to CO_2 is

$$CO + OH \rightarrow CO_2 + H$$

in a system containing oxygen, carbon and hydrogen. Assume that OH and water are in equilibrium with H_2 and O_2 and develop an expression for the rate of CO_2 formation, dependent only upon the concentrations of CO, O_2 and H_2O. (This effectively becomes a global kinetics scheme for CO oxidation.)

4. Calculate the reverse reaction rate for the first reaction in Table 3.3a. Pick a temperature of 700 K and numerically compare the reverse and forward rates. What do you conclude?

4

PREMIXED FLAMES

4.1. INTRODUCTION

A premixed flame is a *propagating combustion wave* in an explosive medium in which the reactants are mixed at the molecular level. Examples are the flame of a laboratory Bunsen burner and the flame in a jet engine afterburner. In the Bunsen burner, fuel (perhaps natural gas) is injected into a tube open at the bottom, allowing air to be drawn in by the fuel jet. The fuel and air mix, coming out of the top of the tube where the mixture can be ignited. In the afterburner, liquid fuel is injected into the hot turbine exhaust gases where the liquid vaporizes and mixes with the gas. This forms a premixed explosive mixture which is ignited farther downstream in the engine.

It is not necessary that only gases are involved. Some propellants in solid propellant rockets consist of molecularly mixed fuel and oxidizer. Some explosives are premixed solids, formed and mixed as liquids and then solidified. Some pure substances are capable of burning (or decomposing) on their own. Examples are ingredients used in rockets such as solid ammonium perchlorate (NH_4ClO_4) or liquid monomethyl hydrazine. Monopropellants are relevant to this chapter because their flames are analytically and physically indistinguishable from premixed flames as described above. Other situations, such as the mixing of solid ingredients macroscopically for making solid propellants, is not, however, considered a premixed situation because mixing at the molecular level is not attained.

4.2. THE HUGONIOT

We now consider a steady, constant area flow, to which the one-dimensional approximation is applied, and we consider no external heat transfer. We presume a chemical, possibly coupled with a normal shock wave, transition between states 1 and 2 as shown in Fig. 4.2a. We introduce here the Mach number as the ratio of the flow velocity to the speed of sound, $M_i = u_i/a_i$ for $i = 1, 2$. The shock wave can only occur, if it occurs at all, in a supersonic stream ($M > 1$). On the other hand, a flame can occur in both supersonic and subsonic ($M < 1$) streams. State 2 in Fig. 4.2a can be either subsonic or supersonic, and we wish to find conditions at 2 given a set of upstream conditions. Under the one-dimensional approximation, the transition from 1 to 2 takes place perpendicular to the oncoming stream. If the upstream flow is subsonic and a flame transition is taking place, we call the process a *deflagration*. If the upstream flow is supersonic we call the transition process a *detonation*.

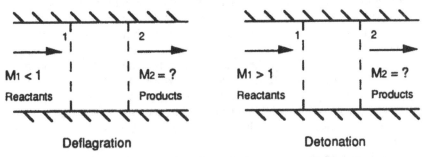

FIGURE 4.2a. Steady premixed combustion wave in a constant area insulated duct.

In flow problems it is most convenient to work with energy quantities on a per unit mass basis. Applying the equations of continuity, momentum, and energy to this flow system[1] we obtain

$$\dot{m} = \rho_1 u_1 = \rho_2 u_2 = \text{constant}, \tag{4.1}$$

$$p_1 + \rho_1 u_1^2 = p_2 + \rho_2 u_2^2, \tag{4.2}$$

$$h_1 + \frac{u_1^2}{2} = h_2 + \frac{u_2^2}{2}. \tag{4.3}$$

Here, $\dot{m}$ is the mass flow rate per unit area and the u's are the flow velocities. The important point to remember in Eq. (4.3) is that the enthalpy contains chemical heats of formation in addition to the sensible enthalpy. Moreover, there may be several species present. It will be presumed that the substance at state 2 is a fluid, either a liquid or a gas. There is no difficulty if the state 1 substance is a solid, as well as either a liquid or gas.

For simplicity at the outset, we assume perfect gases at states 1 and 2, so that

$$p_1 = \frac{\rho_1 R T_1}{W_1} \quad \text{and} \quad p_2 = \frac{\rho_2 R T_2}{W_2}. \tag{4.4}$$

We have from Eq. (2.8) for any species

$$h_i = (h_T - h_{298}^o + (\Delta h_f^o))_i$$

with

$$h = \sum_{i=1}^{M} Y_i h_i$$

at either state 1 or state 2.

For further simplicity we assume *calorically perfect gases* which have constant specific heats, so that the expression for enthalpy becomes

$$h_i = c_{p,i}(T - 298) + (\Delta h_f^o)_{298,i}. \tag{4.5}$$

---

[1] It is assumed that the reader has seen these conservation equations for one-dimensional flows. If some background is necessary, a text such as J. E. John, *Gas Dynamics*. Allyn and Bacon, 1984, Chapts. 4 and 10, is recommended.

When we put this together, the enthalpy at either state 1 or 2 can be represented as

$$h = c_p T + \sum_{i=1}^{M} Y_i [(\Delta h_f^o)_{298,i} - 298 c_{p,i}] \tag{4.6}$$

with an overall specific heat of

$$c_p = \sum_{i=1}^{M} Y_i c_{p,i}.$$

Effectively, what has been done here with the terms $298\, c_{p,i}$ is the transfer of the reference temperature from 298 K to 0 K. The proof of this fact is left to the problems. We may write

$$h = c_p T + \sum_{i=1}^{M} Y_i (\Delta h_f^o)_{0,i}.$$

Evaluating Eq. (4.6) for reactants (state 1) and products (state 2) and placing these in Eq. (4.3), we obtain

$$c_{p,1} T_1 + \frac{u_1^2}{2} + q = c_{p,2} T_2 + \frac{u_2^2}{2} \tag{4.7}$$

with q given by

$$q = \sum_{i=1}^{M} Y_i (\Delta h_f^o)_{0,i} - \sum_{j=1}^{M} Y_j (\Delta h_f^o)_{0,j}. \tag{4.8}$$

The main complication here is the complexity in the evaluation of q; the products and their thermodynamic properties must be known. If the flow problem were such that $u_1 \approx u_2 \approx 0$ and $p_1 \approx p_2$, then Eq. (4.7) would be nothing more or nothing less than a simplified adiabatic flame temperature equation for the temperature T_2.

Example 4.1.
Calculate the q for a gaseous stoichiometric mixture of CO and O_2. Assume the only product is CO_2.

Solution.
From the JANNAF Tables

$$(\Delta h_f^o)_{0,CO_2} = -8.935 \text{ kJ/g}, \qquad (\Delta h_f^o)_{0,CO} = -4.064 \text{ kJ/g}.$$

From the reaction equation

$$CO + \tfrac{1}{2}O_2 \rightarrow CO_2.$$

The mass fractions for the reactants and product are

$$Y_{CO} = 7/11, \qquad Y_{O_2} = 4/11, \qquad Y_{CO_2} = 1.$$

From Eq. (4.8)

$$q = (7/11)(-4.064) + (4/11)(0) - (1)(-8.935) = 6.349 \text{ kJ/g}.$$

A mention of the consequences of the constant c_p assumption should be made here. In combustion, the constant specific heat assumption is usually quite poor, because of the large temperature differences encountered between states 1 and 2. The errors however, show up in poor numbers for answers obtained, not in qualitative behavior of the answers. Since the algebraic complexity is drastically reduced with the constant c_p assumption, we will carry it along here. For further simplicity we will also assume

$$c_{p,1} = c_{p,2} = c_p \qquad \text{and} \qquad W_1 = W_2 = W.$$

These last assumptions are really not too poor if air is the oxidizer in the problem, since usually $f \ll 1$ and nearly inert N_2 is the dominant species in the flow.

We now wish to eliminate some variables. Some perfect gas relations are

$$\gamma \equiv \frac{c_p}{c_v}, \qquad c_p T = \frac{\gamma}{\gamma-1} \frac{p}{\rho}$$

with γ being called the *specific heat ratio*. Using these relations in Eq. (4.7) and manipulating, we obtain

$$q = \frac{\gamma}{\gamma-1} \left(\frac{p_2}{\rho_2} - \frac{p_1}{\rho_1} \right) + \frac{1}{2}(u_2^2 - u_1^2). \qquad (4.9)$$

From Eqs. (4.1) and (4.2)

$$p_2 - p_1 = \rho_1 u_1^2 - \rho_2 u_2^2 = \rho_1 u_1^2 \left(\frac{1}{\rho_1} - \frac{1}{\rho_2} \right)$$

from which u_1 is obtained as

$$u_1^2 = \frac{(p_2 - p_1)}{\rho_1^2 (1/\rho_1 - 1/\rho_2)}. \tag{4.10}$$

When the speed of sound is introduced for calorically perfect gases, $a^2 = \gamma RT/W$, the Mach number is obtained from Eq. (4.10) as

$$M_1^2 = \frac{u_1^2}{a_1^2} = \frac{1}{\gamma} \frac{(p_2/p_1 - 1)}{(1 - \rho_1/\rho_2)}. \tag{4.11}$$

A similar development using Eqs. (4.1) and (4.2) yields the downstream velocity and Mach number as

$$u_2^2 = \frac{1}{\rho_2^2} \frac{(p_2 - p_1)}{(1/\rho_1 - 1/\rho_2)}, \tag{4.12}$$

$$M_2^2 = \frac{1}{\gamma} \frac{(1 - p_1/p_2)}{(\rho_2/\rho_1 - 1)}. \tag{4.13}$$

Placing Eqs. (4.10) and (4.12) in Eq. (4.9) yields a relation between the downstream density and pressure known as the *Hugoniot*:

$$\frac{\gamma}{\gamma - 1} \left(\frac{p_2}{\rho_2} - \frac{p_1}{\rho_1} \right) - \frac{1}{2} (p_2 - p_1) \left(\frac{1}{\rho_1} + \frac{1}{\rho_2} \right) = q. \tag{4.14}$$

This is one of the most important relations in flame and explosion theory. For $q = 0$, which is the case of no chemical heat release, this relation can only be that of a normal shock wave. Such a relation is plotted in Fig. 4.2b. Since expansion shock waves are ruled out by the second law of thermodynamics (it would require an entropy drop in an isolated system), only the high pressure branch is allowed. A normal shock wave is a compression wave which increases the fluid density and pressure. Such a statement is also true for a general substance and is not peculiar to the perfect gas.

Equation (4.14) may now be examined for general q, as is shown in Fig. 4.2c for positive values (exothermic) of q. This figure is split into three branches, *A–B*, *B–C*, and *C–D*. Point *A* is the point at infinity

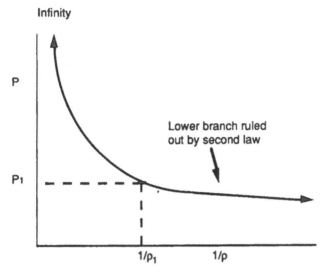

Infinity

P

P₁

Lower branch ruled
out by second law

$1/\rho_1$ $1/\rho$

FIGURE 4.2b. Locations of allowed normal shock wave states (the shock Hugoniot).

in pressure. Sending $p \to \infty$ in Eq. (4.14), for finite p_1 and ρ_1, the density becomes

$$\frac{\rho_2}{\rho_1} \to \frac{\gamma + 1}{\gamma - 1} \qquad (4.15)$$

for any finite q. That is, there is a finite density limit, beyond which the downstream fluid cannot be compressed. The point B is interesting; for $1/\rho_1 = 1/\rho_2$, the offset of the Hugoniot from the initial state is calculated from Eq. (4.14) to be

$$p_2 - p_1 = (\gamma - 1)\rho_1 q.$$

That is, for positive q and all values of γ for real gases ($\gamma > 1$) there is an upward shift of the Hugoniot from the shock Hugoniot.

The B–C branch of Fig. 4.2c is disallowed. For note from Eq. (4.10)

$$(\rho_1 u_1)^2 = \frac{p_2 - p_1}{1/\rho_1 - 1/\rho_2} < 0$$

and the square of a physical quantity cannot be negative. Notice further from Eq. (4.11), that at point A, $M_1(A) \to \infty$, because $p_2 \to \infty$

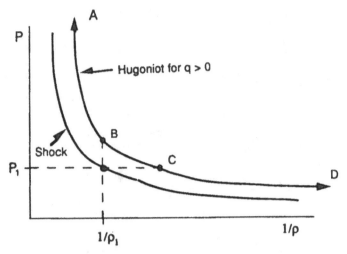

FIGURE 4.2c. The shock Hugoniot ($q = 0$) and the exothermic Hugoniot.

while ρ_2 is finite. Furthermore, from Eq. (4.11), at point B, $M_1(B) \rightarrow \infty$. Since we also know that the shock Hugoniot on the high pressure branch corresponds to only supersonic flow, it is plausible that the complete upper branch corresponds to only supersonic upstream flow.

On the C–D branch of the Hugoniot, similar considerations of Eq. (4.11) yield that $M_1(C) = 0$ and $M_1(D) = 0$. It is consequently plausible that this branch corresponds to subsonic upstream flow. This lower branch is not disallowed as it was in the case of a normal shock wave; further investigation is required on this branch.

From Eq. (4.13), using Eq. (4.15), it is readily shown that

$$M_2(A) < 1, \qquad M_2(B) \rightarrow \infty, \qquad M_2(C) = 0, \qquad M_2(D) \rightarrow \infty.$$

It is consequently plausible that there are $M_2 = 1$ points lying on both the A–B and C–D branches. Now draw the straight lines through the initial point which are tangent to the A–B and C–D branches, as shown in Fig. 4.2d. The tangent points are labeled J and K and are called The Chapman–Jouguet points. It will be stated without proof that indeed these are the states with unit Mach number downstream of the combustion wave (state 2). Consistent with the plausibility arguments above, the A–B branch has upstream Mach number greater than one and the C–D branch has totally subsonic upstream flow. The

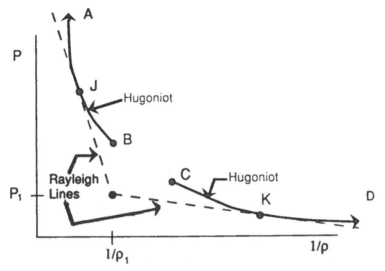

FIGURE 4.2d. Definition of the Chapman–Jouguet points on the Hugoniot.

A–B branch is called the detonation branch of the Hugoniot and the *C–D* branch is called the deflagration branch.

Example 4.2.
For the case of carbon monoxide and oxygen in Example 4.1, calculate the detonation pressure if the initial state is at $p_1 = 1$ bar, $\rho_1 = 1$ kg/m3 and the final state is at $\rho_2 = 3$ kg/m³. Assume that $\gamma = 1.4$.

Solution.

$$P_2\left[\frac{\gamma}{(\gamma-1)\rho_2} - \frac{1}{2}\left(\frac{1}{\rho_1} + \frac{1}{\rho_2}\right)\right]$$

$$= q - P_1\left[\frac{1}{2}\left(\frac{1}{\rho_1} + \frac{1}{\rho_2}\right) - \frac{\gamma}{(\gamma-1)}\left(\frac{1}{\rho_1}\right)\right],$$

$$P_2(0.500) = 6.349 \times 10^6 - (-2.833) \times 10^5$$

$$P_2 = 1.337 \times 10^7 \text{ Nt/m}^2 = 132.7 \text{ bar.}$$

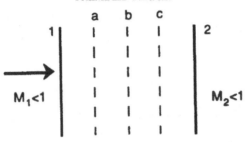

FIGURE 4.2e. Illustration of partial reaction surfaces on the reaction path.

Straight lines drawn through the state 1 point have a special significance. From Eq. (4.10)

$$\dot{m}^2 = \frac{p - p_1}{1/\rho_1 - 1/\rho}$$

is the equation of a straight line on a p versus $1/\rho$ diagram. Allowable straight lines which also intersect the Hugoniot can only have negative slope. The slope magnitude measures the mass flow rate. The maximum flow rate on the deflagration branch occurs at the Chapman–Jouguet point. On the other hand, the minimum mass flow rate occurs at the Chapman–Jouguet point on the detonation branch. These straight lines are called Rayleigh lines and are lines along which mass flow rate and momentum are conserved. Consider now the actual kinetics of the reaction taking place between state 1 and 2. At each station between 1 and 2 a certain fraction of the reaction has been completed. These intermediate states, corresponding to more and more of the product having been formed, are shown as states a, b, and c on Fig. 4.2e.

For the exothermic reaction being considered $(q > 0)$, each intermediate state corresponds to a certain fraction of the overall q having been released. However, corresponding to each fraction of heat release there exits a Hugoniot curve corresponding to that fraction of q. This is shown on Fig. 4.2f on the deflagration branch of the Hugoniot. Drawing the Rayleigh line corresponding to a chosen mass flow and intersecting the "partial heat release" Hugoniots, each intersection is an allowable solution for the intermediate states. These intersections, which satisfy all conservation laws for the given value of the partial heat release, represent intermediate solution points. When state 2 is intersected, reaction is complete.

The drawing of Fig. 4.2f shows intersection on the C–K branch of the deflagration leg of the Hugoniot. The straight line shown consequently intersects all points with a Mach number less than unity. How-

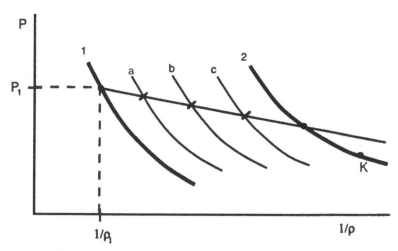

FIGURE 4.2f. Partial reaction Hugoniots and the solution Rayleigh line.

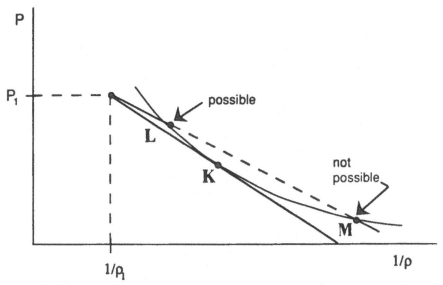

FIGURE 4.2g. Exclusion of supersonic states on the deflagration branch of the Hugoniot.

ever, the situation shown in Fig 4.2g would also seem possible. Point
L is the solution point we have just been discussing. Why is point
M not an allowable solution? It satisfies the Rayleigh line conditions
and the Hugoniot relation. But note, the intermediate states on the
Rayleigh line would have to have a higher q than the reaction allows.
Consequently, the deflagration branch $K-D$ is excluded and only
$C-K$ solutions are allowed. This has as a consequence that subsonic
deflagrations must end up with a subsonic, or at most sonic, speed.
A subsonic to supersonic transition is not allowed along a Rayleigh
line.[2]

There are clearly an infinity of solutions along $C-K$, so which one
corresponds to the real solution with its own mass flow? The real
solution would give us the *flame speed*, which would be unique to the
reactant combination. We will return to this question later; the answer
depends upon the detailed structure of the flame.

Now return to solutions along the detonation branch, recalling that
the Rayleigh line must here have a slope magnitude greater than that
for the line intersecting the Chapman–Jouguet point. It is not possible to argue allowable states without some experimental knowledge
of the structure of a detonation. Consider that a detonation consists
of a shock wave followed by a reaction region. Such a structure is reasonable since a shock wave takes only a few mean distances between
molecules to complete the full transition. Stated otherwise, only a few
molecular collisions are required to make the shock adjustment. On
the other hand, we saw in chemical kinetics that not only collisions
are involved in chemical reactions, but the collisions must have suf-
ficient energy to affect a chemical transition. Moreover, there are
several steps involved in usual flame reactions. The reason that the
shock precedes the reaction, assuming for the moment that the final
state will lie above State J on Fig. 4.2d, is that M_2 is subsonic on the
Hugoniot curve and could not sustain a shock wave; consequently, if
a shock is to be involved, it must come first.

Figure 4.2h shows the situation for a shock followed by a reaction
wave. Point $1'$ is the intersection of the Rayleigh line with the shock
Hugoniot. The flow must be subsonic after traversing the shock. The
flow is also subsonic at state 2 on the Hugoniot, so the reaction path
can be a sequence of states on the Rayleigh line corresponding to
more and more reaction leading to the final state 2. All states are

---

[2]There is a more fundamental reason that such a transition cannot occur. It involves
the second law of thermodynamics, but this explanation will not be discussed here.

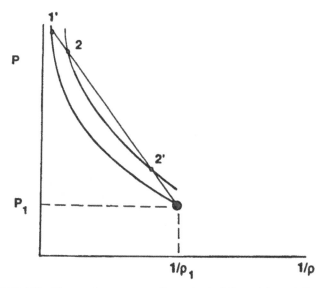

FIGURE 4.2h. Detonation structure for a shock followed by a deflagration.

allowed on the $J–A$ branch of the Hugoniot. State $2'$, which is mathematically possible is, however, ruled out for the same reason as in Fig. 4.2g. All reaction is finished at state 2, and no further heat release is possible along the Rayleigh line to reach point $2'$. The path from $1'$ to 2 is nothing more than a subsonic deflagration following the shock with an initial state of $1'$.

At this point, we have as the only allowable states on the Hugoniot the $J–A$ and $C–K$ states. We still do not know, however, exactly where the final state will lie for a given combustion reaction. Whereas the deflagration requires consideration of combustion wave structure, the detonation will require consideration of boundary conditions. We leave the detonation now, and will return to it later.

4.3. PLANE LAMINAR FLAME

4.3.1. Structure

We now consider the structure of a plane, subsonic combustion wave called the *plane laminar flame*. In this flame, a deflagration, the upstream speed, u, which renders the flame steady in the laboratory frame of reference is called the *laminar flame speed* and is often given

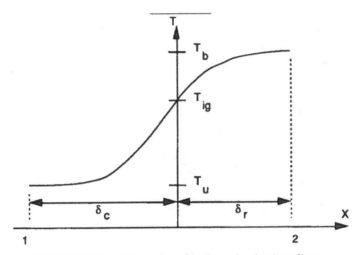

FIGURE 4.3.1a. Thermal profile through a laminar flame.

the symbol S_L. It has the general temperature profile along the flame given in Fig. 4.3.1a. We will call the temperature in the upstream, cold reactants at state 1, T_u, with the "u" standing for "unburned." After completion of reaction, at state 2, the temperature is T_b, with "b" for "burned." The temperature increases monotonically through the wave. Because of the usually large activation energy in the chemical kinetics, most of the reaction takes place in the high temperature portion of the wave. It is therefore convenient to divide the flame onto two regions, one in which the reaction takes place and the other which receives its temperature rise by *thermal conduction* (heat transfer) from the hot zone. The temperature at the point dividing the two zones is the *ignition temperature*, T_{ig}, the temperature at which vigorous reaction starts. We assume the ignition temperature is a property of the reactants in the combustible mixture. We arbitrarily locate the ignition point at $x = 0$.

We proceed to build a theory, called a *thermal theory*, based upon the above picture of the flame structure.[3] The heat transfer will be assumed to be governed by the *Fourier law*, given in one-dimensional form as

$$\bar{q} = -\lambda \frac{dT}{dx} \tag{4.16}$$

---

[3]This theory dates back many years. This picture was put forth first by E. Mallard and H. L. Le Chatelier, *Ann. Mines*, 4 (1883), p. 379.

with $\bar{q}$ the heat transfer rate per unit area. The coefficient is called the *coefficient of thermal conductivity* and is always positive. The negative sign in Eq. (4.16) merely expresses the fact that heat can only flow from a hot to a cold region; $\bar{q}$ is positive in the positive x-direction. The thickness of the conduction zone, between the upstream state and the ignition point, is called δ_c, whereas the thickness of the reaction zone is called δ_r. Equation (4.16) would have to yield $\bar{q}$ at the ignition point as

$$\bar{q} \propto -\lambda \frac{(T_b - T_{\text{ig}})}{\delta_r}$$

since it is dimensionally correct and has the appropriate scaling (that is, if one doubled δ_r, for example and kept the same shape for the temperature curve, the temperature gradient would be halved). For simplicity, therefore, since we are only attempting an approximate theory, we assume

$$\bar{q}_{\text{ig}} = -\lambda \frac{(T_b - T_{\text{ig}})}{\delta_r}. \qquad (4.17)$$

This heat transfer acts to heat up the incoming gases (assumed here). The enthalpy rise to the ignition point must come from this heat transfer, assuming very low speed flow so that kinetic energy is negligible. Therefore,

$$\dot{m}c_p(T_{\text{ig}} - T_u) = -\bar{q}_{\text{ig}}. \qquad (4.18)$$

From Eq. (4.1), $\dot{m} = \rho_u S_L$ so that from Eqs. (4.17) and (4.18)

$$S_L = \left(\frac{\lambda}{\rho_u c_p}\right) \frac{(T_b - T_{\text{ig}})}{(T_{\text{ig}} - T_u)} \frac{1}{\delta_r}. \qquad (4.19)$$

Here δ_r is unknown and will have to be estimated from some other considerations. Moreover, λ has a temperature dependence, albeit weak, so that as evaluated here λ should be evaluated at T_{ig}. However, the assumption will be made that the ignition temperature is sufficiently close to the burned gas temperature that little error is made by evaluating λ at T_b.

Some gas phase thermal conductivities are given in Table 4.3a. Notice that light gases have higher conductivities than heavy gases. Typically, although there is great variability, liquids have conductivities about ten times those for gases, and for solids the multiple is of the order of 10^3–10^4. For gases the conductivity is pressure independent but is approximately proportional to $T^{1/2}$.

TABLE 4.3a
Thermal conductivities of some gases at 273 K

Gas	λ (J/mKs)
H_2	0.174
O_2	0.0240
N_2	0.0237
CH_4	0.0301
He	0.144
Ar	0.0162

We assume global kinetics, whereby the product concentration production rate is given by an expression of the following form:

$$\frac{dc_{pr}}{dt} = R = Ac_f^p c_o^q e^{-E/RT} \qquad (4.20)$$

with c_{pr} being the products concentration, c_f is the local fuel concentration and c_o is the local oxidizer concentration. Since with large E most of the reaction occurs near T_b and there is little density change in this reaction zone

$$\delta_r = u_b \tau_r$$

is a reasonable approximation with τ_r the reaction time. Approximately, the magnitude of an integration of Eq. (4.20) must be

$$c_{pr,b} = R_{av}\tau_r = \tau_r Ac_{f,M}^p c_{o,M}^q e^{-E/RT_b}$$

with R_{av} an average value of the reaction rate. The value of δ_r is, therefore, estimated as

$$\delta_r \approx \frac{u_b c_{pr,b}}{R_{av}} = \frac{\rho_u S_L}{\rho_b} \frac{c_{pr,b}}{R_{av}} = \frac{\rho_u S_L}{W_b R_{av}}. \qquad (4.21)$$

All inaccuracies in the above estimates can be lumped into A. Placing Eq. (4.21) in Eq. (4.19) we have that the final expression for flame speed becomes

$$S_L = \left\{ \left(\frac{\lambda_b}{\rho_u c_p} \right) \frac{(T_b - T_{ig})}{(T_{ig} - T_u)} \frac{W_b}{\rho_u} R_{av} \right\}^{1/2} \qquad (4.22)$$

with Eq. (4.20) for R with $T = T_b$ and the concentrations at their unburned values.

We now explore the behavior of S_L on several of the physical and chemical parameters. First, for large E, the dominant temperature dependence in Eq. (4.22) is through $\exp(-E/RT)$ in $\mathbf{R}$. Since T_b is linear in T_u, the initial reactants' temperature will have a strong effect upon S_L. There are other temperature dependencies in Eq. (4.22) through ρ_u, T_b, and T_u, but they are partially offset by the concentrations in the kinetics expression, since $c_i = pX_i/RT$ and p and q are generally positive. Again, the dominance of the exponential makes the temperature dependence of the result come from the kinetics. This temperature effect is strongly shown in comparison of fuel–oxygen flame speeds to those of a fuel–air speed, where the nitrogen in the air suppresses the flame temperature as compared with that in the fuel–oxygen case.

The second major physical variable in Eq. (4.22) is the pressure. It comes from the concentrations in the reaction rate and the unburned density in Eq. (4.22). The result is

$$S_L \propto p^{(n-2)/2}$$

This pressure dependence is somewhat surprising and counter-intuitive. Since often for hydrocarbon–air flames $n = p + q \approx 1.8$, the laminar flame speed is nearly pressure-independent. The physical reason for this phenomenon is that the flame process is one of mass consumption, but requiring a heat transport process; the mass flow rate is proportional to density (pressure) and the chemical consumption rate is proportional to p^n ($n \approx 2$), but in series with these events is a heat transport process time per unit mass proportional to λ/ρ which slows things down as pressure increases.

The third major effect on flame speed comes from the molecular weight of the reactants. The heat transfer coefficient is roughly proportional to $1/W_b$. That is, light gases have faster random molecular motion and can transmit this by collisions with adjacent molecules (this is what constitutes heat transfer). But the densities of the unburned gases are proportional to W_u. Consequently, in Eq. (4.22) the net effect is

$$S_L \propto \frac{1}{W_u}.$$

Light gases have the highest flame speeds; for this reason the hydrogen–oxygen system has one of the highest flame speeds known, but the speeds of hydrocarbon–oxygen systems are generally quite low.

Typical numbers for hydrocarbon–air flames are that flame speeds are of the order of magnitude of tens of cm/s, flame thicknesses

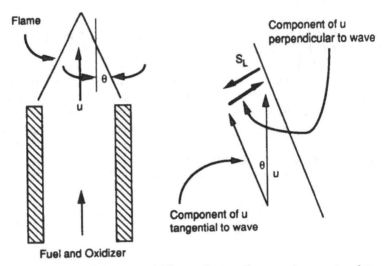

FIGURE 4.3.2a. Schematic of Bunsen burner flame and geometry determining S_L.

$(\delta_c + \delta_r)$ are around a few mm and reaction times are near a few ms. At these speeds, viewing Eqs. (4.2) and (4.3), the kinetic energies of the flow are negligible compared to the thermal and chemical energies, and the pressure is nearly constant. As a consequence, the temperature attained in these flames is very close to the adiabatic flame temperature. Moreover, on the Hugoniot, the final point lies very close to point C (see Fig. 4.2c).

4.3.2. Experimental Results

In order to test the above predictions and assess their accuracy, we now look at some of the experimental results on laminar flames. Moreover, it appears clear that it may be possible to measure some of the parameters in Eq. (4.22) by measuring flame speeds and their dependence on various parameters. One of the most popular methods of measurement of flame speeds is through the simple Bunsen burner, although there are many sources of inaccuracies in the method. Shown in Fig. 4.3.2a is a schematic of a laminar flame sitting on top of a Bunsen burner. In order for the flame to be stationary with respect to the burner it must propagate into the unburned gases with a component of flame speed in the direction of the oncoming gases that balances the vertical gas flow speed. In general, the flow speed is higher than the flame speed. The only way to accommodate this situation is for the flame to sit at an angle to the oncoming stream. Notice that if the flame speed were greater than the gas speed, the flame would propagate into the burner tube and we could not see it. Such a phenomenon is called *flashback* and will be covered later. From the diagram in Fig. 4.3.2a, the flame speed is given by $S_L = u\sin(\theta)$.

There are several problems in determining the flame speed by the Bunsen burner method, but it gives a fast, easy way of at least getting the correct magnitude of the flame speed. The first problem is that the flame is not a straight cone, and the position of measurement affects the answer. Secondly, the oncoming magnitude and direction of the oncoming flow velocity is not known with precision, usually. Thirdly, the optical method of determining the angle (visual or by shadowgraph, Schlieren or interferometric photography) will affect the result because the apparent cone angle changes with the method. Finally, there are always some effects of the burner wall. Having mentioned these troubles, many of the literature results on flame speeds have nevertheless been obtained with this method.

Consider first the effect of pressure on the laminar flame speed. Figure 4.3.2b shows some results for three hydrocarbons burning with air.[4] For the ethylene-air case notice the extreme insensitivity of the flame speed with pressure; the flame speed only drops by about 40% for a factor of 100 change in the pressure. Results such as those shown in Fig. 4.3.2b can be used to estimate the overall order of the effective, global reaction. From the above theory

$$\ln(S_L) = \text{constant} + \left(\frac{n-2}{2}\right)\ln(p)$$

so that the local slope of the flame speed vs. pressure plot determines n (which varies somewhat with pressure). One inaccuracy here is that there is a mild effect of pressure on the flame temperature, because of dissociation. Consequently, the exponential in the flame speed formula also depends somewhat on pressure. This effect could be corrected for, if the activation energy were known and the adiabatic flame temperature calculations were made.

The effect of equivalence ratio is dramatic, as shown in Fig. 4.3.2c.[5] The flame temperature maximizes near $\phi = 1$ so that the reaction rate

---

[4]From I. Glassman, *Combustion*. Orlando: Academic Press, 1987, p. 136.

[5]From I. Glassman, *Combustion*. Orlando: Academic Press, 1987, p. 137.

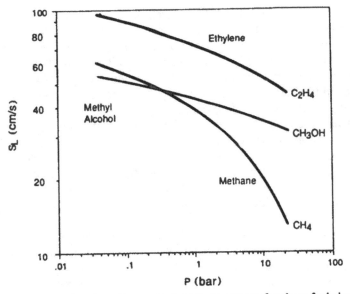

FIGURE 4.3.2b. Laminar flame speed against pressure for three fuels in sto-ichiometric mixtures with air for an initial temperature of 298 K.

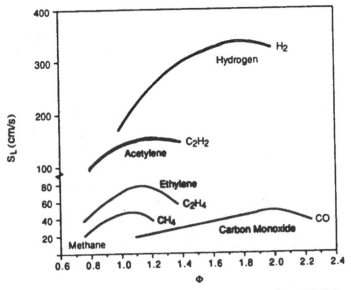

FIGURE 4.3.2c. Laminar flame speed at $p = 1$ bar and $T_u = 298$ K for five fuels burning with air.

exponential maximizes here and the flame speed is near a maximum. However, for almost all cases the flame speed maximum occurs somewhat to the right of the stoichiometric mixture point. This is often due to a molecular weight effect. Notice the hydrogen–air case in Fig. 4.3.2c. When the mixture ratio is hydrogen rich there is left over hydrogen in the reaction. Since hydrogen has low molecular weight and $S_L \propto W_u^{-1}$ the maximum speed is on the fuel rich side of $\phi = 1$. For hydrocarbons, excess fuel usually means a lowered effective value of c_p. From Eq. (4.22), such an effect tends to increase the flame speed slightly and causes a flame speed peak on the fuel rich side. The carbon monoxide result in Fig. 4.3.2c is unique and is due to a complex chemical kinetics cause which we will not discuss.

The adiabatic flame temperatures for the mixtures shown in Fig. 4.3.2c are roughly all the same for hydrogen, acetylene, and carbon monoxide. Nevertheless, the flame speeds are markedly different. More than just a molecular weight effect, these differences are due to different chemical kinetics laws for the different substances. Not only is the pre-exponential factor different for the different chemicals but so is the effective value of the activation energy.

The effects of molecular weight and heat capacity are shown in Fig. 4.3.2d.[6] The method used here is to substitute inert noble gases for nitrogen in air. The noble gases are monatomic gases, whereas the nitrogen is diatomic. For Ar and He the molar heat capacity is 2.5 R. For nitrogen, depending the temperature, C_p lies between 3.5 and 4.5 times the universal gas constant, depending upon the temperature. Viewing the energy equation, Eq. (4.18), we may write it on a molar basis as well as the per unit mass basis. That is,

$$c_p(T_b - T_u) = q \qquad \text{and} \qquad C_p(T_b - T_u) = Q$$

are equivalent, with Q being qW. For all of the flames of Fig. 4.3.2d, the Q value is the same, being independent of the inert being used. But with Ar and He the heat capacity is down, as compared with the nitrogen case. Consequently, the flame temperature is up, the kinetics are faster, and the flame speed is higher. Comparing the He case to the Ar case, He has a substantially lower molecular weight than Ar. Low molecular weight means higher flame speed, as described above in connection with Eq. (4.22).

[6] W. H. Clingman, Jr., R. S. Brokaw, and R. D. Pease, *Fourth Symposium (Internat.) on Combustion*. Baltimore: Williams and Wilkins, MD, 1953, p. 310.

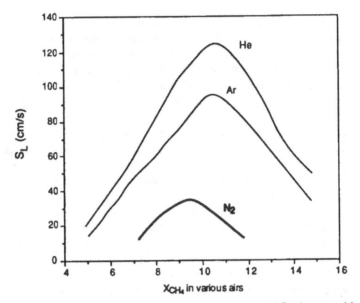

FIGURE 4.3.2d. Laminar flame speed of methane in "air" mixtures with argon and helium substituted for nitrogen.

Example 4.3.
For stoichiometric methane and "air" made up of 21% O_2 and 79% Ar, calculate the ratio of the laminar flame speed to its value if the Ar were replaced by He.

Solution.
Here the only change is in the reactant molecular weight. The reaction equation

$$CH_4 + 2\left(O_2 + \frac{0.79}{0.21}\text{Inert}\right) \rightarrow CO_2 + 2H_2O + 7.52 \text{ Inert.}$$

The molecular weight in the first case is

$$W_u(\text{Ar}) = \frac{16 + 2 \cdot 32 + 2 \cdot 3.76 \cdot 40}{1 + 2 + 2 \cdot 3.76} = 36.2 \text{ g/mol}$$

while

$$W_u(\text{He}) = 9.03 \text{ g/mol.}$$

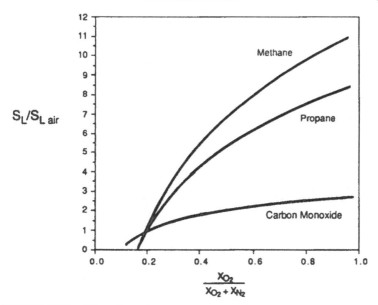

FIGURE 4.3.2e. The effect of oxygen concentration on flame speed relative to the flame speed using air.

Therefore,

$$S_L(\text{Ar})/S_L(\text{He}) = 9.03/36.2 = 0.25.$$

The actual result from Fig. 4.3.2d does not exhibit this drastic a change.

Figure 4.3.2e illustrates a complex chemical kinetics effect.[7] Shown is the ratio of the flame speed of the fuel in air to the flame speed of the fuel in various mixtures of oxygen and nitrogen. Of course, the higher the oxygen level the higher the adiabatic flame temperature and the flame has a higher speed. Comparing the methane results to those for carbon monoxide, there is a drastically different change in flame speed with oxygen level for the two fuels. On the other hand, the change in flame temperature is about the same for both; the temperatures for CO oxidation are about 2400 K and 3220 K in air and pure oxygen, respectively. For CH_4 these numbers are 2210 K and 3030 K. What this implies are quite different effective activation energies for the global kinetics between the two fuels. Moreover, the

---

[7]K. S. Zebatakis, *U.S. Bureau of Mines Bulletin* No. 627, 1965.

TABLE 4.3.3a
Flammability limits in air and oxygen at 1 bar and 298 K
(Mole fraction of fuel)

Fuel	Lean Air	Lean Oxygen	Rich Air	Rich Oxygen
Hydrogen	0.04	0.04	0.75	0.94
Carbon monoxide	0.12	0.16	0.74	0.94
Methane	0.05	0.05	0.15	0.61
Propane	0.02	0.02	0.10	0.55

rate dependence on oxygen concentration can be different for the two cases. Also, the pre-exponential factor in global kinetics can change as the mixture changes. Global kinetics laws are often too oversimplified to trus: detailed flame calculations.

Not covered here is the effect of ignition temperature. Nor has it been indicated how to choose it. This question will be returned to in the chapter on ignition.

4.3.3. Flammability Limits, Flashback and Quenching

There are some fuel-oxidizer mixtures that will not propagate a flame regardless of the strength of the ignition source. Even though a theoretical calculation of a flame temperature can be made, and it may be quite high, the flame cannot sustain itself in a self-propagating mode. Although the explanation is by no means precise, this situation may be thought of as a case in which the ignition temperature is higher than the flame temperature. The ignition temperature is a chemical kinetics concept while the flame temperature comes from the thermodynamics. If the kinetics are too slow at the maximum temperature the system can achieve, then the flame will go out. This is by no means, however, the only explanation for the existence of *flammability limits*.

The word "limit" refers to the fuel rich and fuel lean mixture ratios outside of which a flame will not self sustain. These are shown for several fuels with both air and oxygen as the oxidizers in Table 4.3.3a. These limits are shown for the conditions of 1 bar and 298 K. The limits can be widened if the mixture is heated to a higher temperature or if the pressure is raised. Both of those actions increase chemical kinetic rates. However, the limits also depend upon the confining material and the proximity of walls; but these effects are discussed later in connection with *quenching*. Interestingly, the flammability limits also depend upon gravity; an upward propagating flame usually has dif-

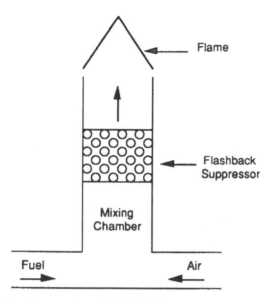

FIGURE 4.3.3a. Suppression of flashback in a premixed combustion system.

ferent limits than a downward propagation one. This is of extreme interest, because spacecraft fire safety is one area of interest in combustion science. Other areas which demand knowledge of flammability concern accidental explosions in coal mines, grain elevators, homes and laboratories.

A rough rule of thumb is that the lean flammability limit occurs at about 50% of the fuel at stoichiometric conditions, while the rich limit occurs at about three times the stoichiometric fuel. The lean limits are the same in both air and oxygen, because there is more than enough oxygen to burn the fuel; the excess oxygen is diluent and behaves just as the nitrogen in the air. However, the rich limit depends strongly upon the oxygen concentration, because the flame temperature is strongly dependent upon this concentration when fuel-rich.

If we have a premixed explosive mixture for which the laminar flame speed is greater than the flow velocity inside of a tube, any flame generated will propagate into the tube. That is, the flame will flashback. This can be a dangerous situation. In laboratory practice a way to counter this event is to make a flashback suppressor, as shown in Fig. 4.3.3a. Filling a portion of the tube with, say, steel balls or steel wool does two things. First, the constriction raises the velocity of the gases, against which the flame is propagating. Raising this

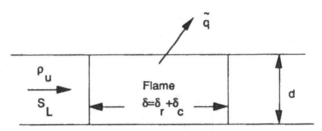

FIGURE 4.3.3b. Geometry of flame quenching by heat conduction.

speed above the laminar flame speed will stop any further propagation. Secondly, the tube fill material acts as an increased heat transfer area. The flame will become non-adiabatic, drop its temperature and, hopefully, go out. Another effect here is that the increased solid surface area will act as an absorption sink for flame radicals, aiding in the extinguishment process. This process of flame extinguishment is called quenching. Once again, while not precise, this may be thought of as lowering the flame temperature below the ignition temperature.

Consider flame propagation between two parallel walls as shown in Fig. 4.3.3b. If the ratio of heat loss rate across the flame thickness, δ, to the heat generation rate in the flame is greater than some critical number, η, it is reasonable that the flame will extinguish. That is, a criterion for the flame to go out is

$$\frac{(2\tilde{q}\delta)}{(\rho_u S_L q d)} \geq \eta.$$

But we already have and $\tilde{q} \propto \lambda$ and $\delta \propto \rho_u S_L/\mathbf{R}$ so that the quenching criterion is

$$d_p \equiv (\text{quenching distance}) \propto \frac{\lambda}{\mathbf{R}}.$$

From Eq. (4.22) we have

$$S_L^2 \propto \frac{\lambda \mathbf{R}}{(\rho_u c_p)} \qquad \text{and} \qquad \frac{\lambda}{\rho_u} \propto \frac{\lambda}{p}.$$

The quenching criterion may therefore be written

$$d_p \propto \frac{\lambda^2}{(p S_L^2)}. \tag{4.23}$$

This relation will also be useful in the later study of spark ignition. The behavior is schematically shown in Fig. 4.3.3c. It should also be

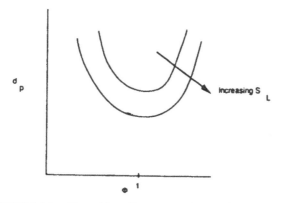

FIGURE 4.3.3c. Quenching diameter against equivalence ratio.

noted that absorption of radicals at the wall follows a rate law proportional to the pressure. Consequently, independent of the thermal quenching of Eq. (4.23), such a mechanism of quenching would behave with pressure as in Eq. (4.23). A typical plot of quenching diameter for cylindrical tubes is shown in Fig. 4.3.3d, showing the expected behavior with mixture ratio.[8]

4.4. HETEROGENEOUS PLANE DEFLAGRATIONS

Some solid rocket propellants are mixed at the molecular level, and, with some modification, can be analytically treated as in Section 4.3.1. An example would be so-called double base propellants made from nitrocellulose and nitroglycerin. The thermal profile is shown in Fig. 4.4a. The dominant difference as compared with Fig. 4.3 is the break in temperature slope at the solid–gas interface. The solid usually requires some heat input to gasify it and this heat is the *heat of pyrolysis*, or the heat of gasification, or the *latent heat of sublimation*. Consequently, the gas phase heat transfer at the interface goes toward providing both the latent heat and continued heat transfer into the solid. An energy balance at the ignition point, whereby all heat transfer from the reaction zone goes toward providing (a) the solid heat-up to the solid–gas interface, (b) the latent heat and (c) gas phase heat-up

---

[8]B. Lewis and Guenther von Elba, *Combustion, Flames and Explosions of Gases*. New York: Academic Press, 1961, p. 230.

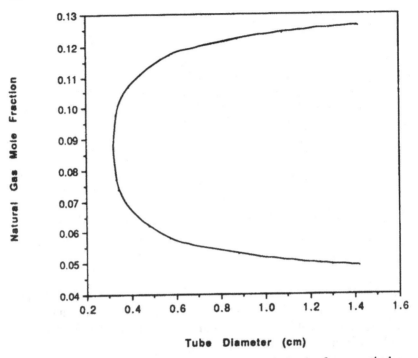

FIGURE 4.3.3d. Quenching diameter in a cylindrical tube for a particular natural gas.

to the ignition point becomes

$$\rho_u u_u [c_{p,s}(T_s - T_u) + L + c_{p,g}(T_{ig} - T_s)] = \frac{\lambda(T_b - T_{ig})}{\delta_r}. \quad (4.24)$$

All other considerations pertinent to the gas phase laminar flame hold here, but it must be borne in mind that the upstream density is that of a solid. In solid propellant flames, the flame speed is called the *deflagration rate* and is usually given the symbol r. From the same considerations leading to Eq. (4.22) we have

$$\delta_r \approx \frac{u_b c_{pr,b}}{\mathbf{R}} = \frac{\rho_u u_u}{W_b \mathbf{R}},$$

$$r = \left\{ \frac{\lambda_b (T_b - T_{ig}) W_b \mathbf{R}}{\rho_u^2 c_{p,s} \left[(T_s - T_u) + \dfrac{L}{c_{p,s}} + \left(\dfrac{c_{p,g}}{c_{p,s}} \right)(T_{ig} - T_s) \right]} \right\}^{1/2}. \quad (4.25)$$

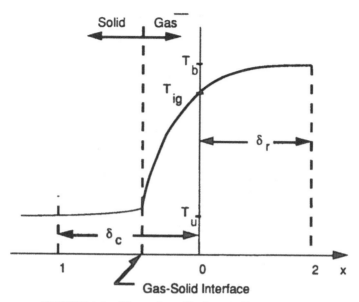

FIGURE 4.4a. Thermal profile for a deflagrating solid.

Because the solid phase density does not scale with pressure, the pressure dependence of the deflagration rate is different compared with that of a purely gas phase flame. Whereas the gas flame speed was usually nearly independent of pressure, the solid propellant experiences a rate proportional to $p^{n/2}$, which is a reasonably strong dependence.

4.5. PLANE DETONATIONS

Referring back to Fig. 4.2h, we return to the detonation branch of the Hugoniot curve. Once again, this combustion wave propagates at a speed which is supersonic with respect to the oncoming flow. Moreover, the only solution possible lies on the A–J branch of the Hugoniot which has the speed of the downstream gases at most sonic with respect to the wave. Uses for the detonation wave, which can have extremely high downstream pressures, include conventional warheads, shaped charges and the initiating explosion for the fission bomb which in turn is the initiator of the fusion bomb. Peaceful uses include blasts for mining and construction or demolition.

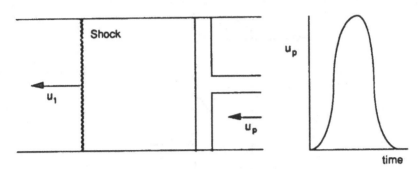

FIGURE 4.5a. Detonation initiation by a piston with impulsive loading.

Given an explosive mixture capable of detonation, one way of initiation is to start a strong shock wave by an impulsive loading on the mixture, as shown in Fig. 4.4a. We presume the shock is strong enough to heat the trailing gas to the ignition temperature and start the reaction. In Fig. 4.2h the process path goes $1 - 1' - 2$. If we pick up the tube in Fig. 4.5a and translate it to the right at speed u, the wave is now stationary with respect to a fixed observer and Eqs. (4.1)–(4.3) apply. Forming the product $\dot{m}(1/\rho_1 - 1/\rho_2)$, we note this must be positive on the detonation branch (and negative on the deflagration branch) of the Hugoniot. But

$$\dot{m}\left(\frac{1}{\rho_1} - \frac{1}{\rho_2}\right) = \left(\frac{\rho_1 u_1}{\rho_1} - \frac{\rho_2 u_2}{\rho_2}\right) = u_1 - u_2 \quad \begin{array}{l} > 0 \quad \text{for detonation} \\ < 0 \quad \text{for deflagration.} \end{array}$$

That is, the gases slow down in going through a detonation. Returning to the unsteady frame of reference in Fig. 4.5a, the piston is stationary after the pulse is given, but the gases must be attempting to pull away from the piston face. The detonation is trying to "drag" the gas from the piston and cause a vacuum. On the other hand, we know that any disturbance generated by the piston can travel at the speed of sound in the trailing gases and reach the wave.

The disturbance here is an expansion wave generated by the stationary piston. This travels to the detonation wave and tries to lower the pressure downstream of the wave. The detonation can accommodate this situation by slowing down, creating a weaker pressure rise across it. The result is that the final state attempts to move to point J—the Chapman–Jouguet point. In fact, at J the downstream gases are just sonic relative to the wave. Beyond this point no disturbance from the piston can catch up with the wave; this, then, is the equilibrium situation for the wave if the piston is stationary.

It should be clear, however, if the piston were given some steady speed at the end of the pulse phase, that any detonation on the A–J branch can be maintained. The piston speed must simply match the gas speed demanded by the detonation solution. The most common situation encountered in practice, however, is the Chapman–Jouguet solution, because everything surrounding the detonation process is usually at rest.

There are some quite simple relations which emerge for the Chapman–Jouguet detonation. Setting $u_2 = a_2$, assuming $p_2 \gg p_1$ and using perfect gas relations, Eq. (4.11) yields

$$(\rho_2 u_2)^2 = \frac{p_2 - p_1}{1/\rho_1 - 1/\rho_2} \approx \frac{p_2}{1/\rho_1 - 1/\rho_2}$$

$$= (\rho_2 a_2)^2 = \frac{\rho_2^2 \gamma p_2}{\rho_2}.$$

Cancelling p_2 yields in this limit,

$$\frac{\rho_2}{\rho_1} = \frac{\gamma + 1}{\gamma}. \tag{4.26}$$

Note that this is not the same as previously derived for the density limit in the limit of infinite pressure on the Hugoniot. Placing Eq. (4.26) in Eq. (4.14) and again assuming $p_2 \gg p_1$ some tedious algebra yields

$$p_2 = 2(\gamma - 1)\rho_1 q. \tag{4.27}$$

Then from Eqs. (4.10) and (4.27), under the same approximations, the Chapman–Jouguet detonation speed is

$$u_1 = [2q(\gamma^2 - 1)]^{1/2}. \tag{4.28}$$

At an intermediate step in derivation of Eq. (4.28), if the perfect gas law $p = RT/W$ is used, an alternate expression for the detonation speed is

$$u_1 = (\gamma + 1)\left(\frac{RT_2}{\gamma W}\right)^{1/2}. \tag{4.29}$$

TABLE 4.5a
Detonation parameters for several hydrogen–oxygen–additive mixtures[1]
$P_1 = 1$ bar, $T_1 = 291$ K
(Theoretical P_2 and T_2, experimental u_1)

Mixture	P_2 (bar)	T_2 (K)	u_1 (m/s)
X	18.05	3583	2819
$X + 5\text{He}$	16.32	3097	3160
$X + 5\text{Ar}$	16.32	3097	1700
$X + 5\text{H}_2$	15.97	2975	3527
$X + 5\text{O}_2$	14.13	2620	1700
$X + 5\text{N}_2$	14.39	2685	1822

[1]This is taken from I. Glassman, *Combustion*, 2nd ed. Orlando: Academic Press, 1987, p. 220, but there are several sources for the original data. Glassman abstracted the data from B. Lewis and G. von Elbe, *Combustion, Flames and Explosions of Gases*, 2nd ed. New York: Academic Press, 1961. The first abstraction of the original data is in a nearly inaccessible report and is quoted here in G. B. Kistiakowski, E. B. Wilson, Jr. and R. S. Halford, in *Underwater Explosion Research*, Vol. 1, Office of Naval Research, 1950, p. 209. Some of the original data are in B. Lewis and J. Friauf, *J. Amer. Chem. Soc.* 52 (1930), p. 3905, and H. B. Dixon, *Trans. Roy Soc. London*, A126 (1893), p. 97. The theoretical calculations were made based upon incomplete thermodynamic data and before modern computational techniques made equilibrium calculations routine. Nevertheless, there is reasonable agreement between theory and experiment and the data serve to illustrate the desired points.

In order to explore the various parameters entering into the detonation formulas we consider the hydrogen–oxygen detonation. Table 4.5a shows some detonation results abstracted from several sources for the stoichiometric mixture of hydrogen–oxygen with several additives. We call the basic mixture $X = 2\text{H}_2 + \text{O}_2$. To it we add several diluents or fuel or oxidizer. Shown in the table are calculated pressure, temperature and measured Chapman–Jouguet velocity.

In reasoning the effects of the various additives to X, we appeal to Eqs. (4.27)–(4.29) and Eq. (4.8) as well as to the JANNAF Tables for the thermodynamic properties of the additives. One additional equation which aids in the discussion is obtained by placing the perfect gas law in Eq. (4.27) and using Eq. (4.26) to obtain T_2 as

$$T_2 = \frac{\gamma(\gamma - 1)}{\gamma + 1} \frac{qW_2}{R}. \tag{4.30}$$

In all of the cases above the dominant exothermic product of the reaction is water. Notice in Eqs. (4.27) and (4.30) that multiplication of q by either ρ_1 or W_2 turns the heat release into a volumetric or molar heat release which is the same for addition of either He or Ar in equal volume (moles). For that reason, in Table 4.5a, the pressure and temperature are the same for He or Ar addition. In both cases, however, the heat release per mixture mole is down, because of dilution, accounting for the lowered temperature compared with the baseline case. Nevertheless, the detonation speed is higher for the case of He addition as compared with the baseline case. This is an effect of low W in Eq. (4.29). With the higher W for Ar, the detonation speed is indeed down from the pure $H_2 - O_2$ case. There are also effects of γ in Eqs. (4.27)–(4.30) which tend to elevate p_2, T_2 and u_1 with He and Ar addition because the $\gamma = 1.67$ for these gases.

Turning now to the effect of excess H_2 addition, the H_2 acts as a diluent, lowering p_2 and T_2 compared with the base case. Surprisingly, however, the detonation speed goes up because of the strong effect of the lowered molecular weight in Eq. (4.29). This is basically a sound speed effect, whereby the sound speed rises with molecular weight drop. In a Chapman–Jouquet detonation the rise in the downstream speed (sound speed) demands a faster upstream speed (demanded by continuity since the density ratio is fixed).

The effect of N_2 and O_2 dilution in Table 4.5a is as expected—a drop in p_2 and T_2 and in detonation speed. There is, however, a slight difference in results depending upon whether O_2 or N_2 is used. Part of the difference comes from the fact that O_2 has a slightly higher molecular weight than does N_2. Another effect comes from the fact that the O_2 bond is much easier to break than the N_2 bond (see Table 2.3a). Thus, O_2 dissociates much more in the products of combustion, lowering the temperature and detonation speed.

The results of Table 4.5a are for gases and, if starting out at room temperature and pressure, the detonation pressures are of the order of 10 to 20 bar. Notice, however, the effect of ρ_1 in Eq. (4.27). That equation is also valid for solid explosives which generally have densities of the order of 1000 times greater than those of gases. For comparable heat release we are speaking of detonation pressures around 100 kbar! Hence, the usefulness of the detonation.

Finally, it should be mentioned that there are *detonability limits* just as with flammability limits. The two are not the same, but the detonation limits will not be covered in this text. Other complications with detonations are that the actual internal structure is complex and is three-dimensional. This does not invalidate the formulas developed by one-dimensional theory, because the overall discontinuity analysis must still be preserved if the beginning and end states are one-dimensional, as they are to good approximation.

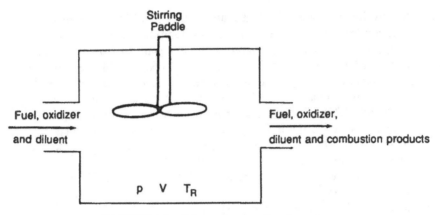

FIGURE 4.6a. Schematic of a stirred reactor.

Example 4.4.

For the case of Example 4.2, calculate the Chapman–Jouguet detonation speed. Here the downstream density is not given as in Example 4.2, but is a consequence of the Chapman–Jouguet assumption.

Solution.

From Eq. (4.28) and the γ and q from Examples 4.1 and 4.2,

$$u_1 = [2q(\gamma^2 - 1)]^{1/2} = [2(6.349) \times 10^6(1.4^2 - 1)]^{1/2} = 3400 \text{ m/s}.$$

This number is high because no dissociation has been in the products.

4.6. THE STIRRED REACTOR

We now ask the question as to the maximum rate of possible conversion of reactants to products in a practical flow device of limited size. This is of immense technological interest, because there are usually space restrictions on real combustion chambers, as in an airplane, for example. This section departs somewhat from purely premixed systems, since the question really asked is what happens when initially unmixed reactants are put into a combustion chamber, then mixed and then reacted before discharge? However, we are going to presume a situation in which the mixing is so fast that for all practical purposes the composition is everywhere uniform so that the actual combustion process is in a premixed system.

Conceptually we consider a *stirred reactor* as shown schematically in Fig. 4.6a. Fuel plus oxidizer and diluent are injected into a chamber and, with chemical reaction occurring, are violently stirred.[9] The stirring process is so strong that there is no combustion wave, but reaction occurs throughout the vessel, and products of combustion are also stirred into the reactants. That is, we imagine the composition is uniform throughout the vessel, consisting of reactants and products. If the mass flow is small there is a long residence time in the vessel and we may imagine that nearly complete reaction occurs. The opposite is true if the mass flow is too large. In fact, we may reason that if the mass flow is too large and no reaction occurs, the flame will *blow out*. It is precisely this blowout condition that is of interest.

Several assumptions will be made in this analysis. First, the stirring is so strong that everything is perfectly mixed and the only process limiting the consumption rate of reactants is chemical kinetics. Second, we assume that thermally and calorically perfect gases with a uniform c_p. Third, we consider only steady flow with a uniform pressure and temperature, T_R, in the vessel, and, finally, we assume that velocities are so low that kinetic energies are negligible in comparison with thermal energies.

The chemical reaction considered is

$$f \text{ Fuel} + o \text{ Oxidizer} + d \text{ Diluent} \rightarrow p \text{ Products} + d \text{ Diluent}$$

where f and o are in stoichiometric proportions and the diluent may contain excess oxidizer or excess fuel. We define the extent of reaction, ξ, as

$$\xi = 1 - \frac{Y_{F,\text{out}}}{Y_{F,\text{in}}}.$$

Clearly, $0 < \xi < 1$. Under the stated assumptions the enthalpy rise is

$$\dot{m}(h_{\text{out}} - h_{\text{in}}) = \dot{m}c_p(T_R - T_{\text{in}}) = \dot{m}c_p\xi(T_{\text{ad}} - T_{\text{in}})$$

since the reactor temperature is linear in ξ. That is,

$$T_R = T_{\text{in}} + \xi(T_{\text{ad}} - T_{\text{in}}).$$

---

[9]To close approximation the stirred reactor condition can be achieved, but not as shown in Fig. 4.6a. See J. E. Nenninger, A. Kridiotis, I. Chomiak, J. P. Longwell, and A. F. Sarafim, *Twentieth Symposium (Internat.) on Combustion*, The Combustion Institute, 1985, p. 473.

Let w_F be the production rate of the fuel per unit vessel volume, V. It follows that

$$m_{F,\text{out}} - m_{F,\text{in}} = w_F V = \dot{m}(Y_{F,\text{out}} - Y_{F,\text{in}}) = -\dot{m}\xi Y_{F,\text{in}}.$$

Finally, we introduce a global kinetics law of the following form:

$$w_F = W_F \frac{dc_F}{dt} = -W_F A c_F^p c_O^q e^{-E/RT_R}$$

and introduce this into the prior equation, using the perfect gas law. By construction $X_O = X_F(o/f)$, we have the overall reaction order $n = p + q$ and recall $X_i = WY_i/W_i$. Collecting and forming the *mass loading parameter*, $\dot{m}/V\,p^n$, the final result is

$$\frac{\dot{m}}{V p^n} = \left(\frac{W_F}{(RT_R)^n}\right) A e^{-E/RT_R} \left(\frac{W_F^p W_0^q}{W_{\text{out}}^n}\right) Y_{F,\text{in}}^{n-1} \left(\frac{(1-\xi)^n}{\xi}\right).$$

$$(4.31)$$

Assuming W_{out} is only a weak function of ξ, the qualitative result of Eq. (4.31) is shown in Fig. 4.6b.

Figure 4.6b shows three solution branches, labeled as such. Branch 1, a low temperature branch, is valid if the chemical kinetics law is valid. However, the kinetics law is generally developed for high temperature reactions (such as in a laminar flame) and is not valid on Branch 1. Consequently, Branch 1 is never observed. Branch 2 is physically unappealing since it requires that as the mass flow rate is increased, the amount of reaction (the temperature) goes up. In fact, a stability analysis (unsteady analysis) shows that Branch 2 is not stable; the true solution would jump to either Branch 1 or Branch 3. The actual solution lies on Branch 3 and it says there is a maximum mass loading parameter that can be tolerated. Any attempts to increase the mass flow beyond the maximum will force the flame to blow out. The temperature at which this occurs is, of course, lower than the adiabatic flame temperature, because the reaction is incomplete.

Differentiating Eq. (4.31) by ξ (recalling that T_R is a function of ξ) and setting the result equal to zero will yield an extremum (actually, two) in the mass loading parameter. The maximum will yield the value for an optimum value of $\xi = \xi_{\text{opt}}$ for the maximum value of the mass loading parameter. However, a closed form solution is not possible. As an exercise in approximation theory the student should perform the following operations:

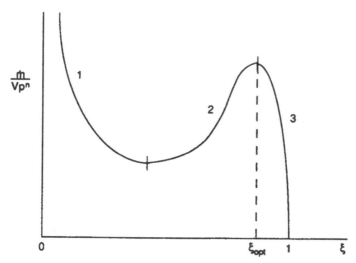

FIGURE 4.6b. Mass loading parameter as a function of the extent of reaction.

Assume $\xi_{opt} = 1 - \epsilon$, with $\epsilon \ll 1$.

As a consequence, T_R is approximately T_{ad}. The consequence, using truncated Taylor series expansions, will be that

$$\xi_{opt} = 1 - 1/(E/RT_{ad} - 2n).$$

Typically, ξ_{opt} is approximately about 0.7 with T_R 70% of T_{ad}.

Because of the assumptions of perfect mixing and a global kinetics law, both of which may be flawed in real situation, the quantitative use of the above theory is often unjustified. However, the qualitative result is instructive and reasonable. Later, in connection with flame stabilization discussions, it will be seen that the stirred reactor is a reasonable model for certain portions of a real combustion chamber. However, for good fuel utilization we do not want to exhaust unburned fuel from a combustor. Consequently, real combustion chambers operate much closer to the complete combustion limit with $T \approx T_{ad}$. This requires more combustor volume than would occur near the maximum mass loading parameter. Typically, the actual mass loading parameter for real combustion chambers is a factor of 1/10 to 1/100 that of the maximum attainable loading.

4.7. PROBLEMS

1. For the reactive system $H_2 + F_2 \rightarrow 2HF$, calculate the q for the Hugoniot curve. Assuming thermally and calorically perfect HF and using the thermodynamic properties of the HF, plot the Hugoniot curve. Calculate the Chapman–Jouguet detonation speed. (Answer: $q = 13.6$ kJ/g)

2. A slab of solid propellant is deflagrating at a pressure of 20 bar. The overall order of the gas phase reaction is 1.5. If the pressure is raised to 30 bar, what is the new regression rate? what assumptions are made to make this calculation?

3. Consider gaseous methane burning with air in a deflagration. From Fig. 4.3.2b estimate the order of the reaction near 1 bar pressure. Assume $p = q$, $E = 170$ kJ/mol, an ignition temperature of 1000 K and a thermal conductivity of 5×10^{-4} J/(cm K s). Assuming $c_p = 1$ J/(g K), a stoichiometric mixture and an initial temperature of 300 K, calculate the factor A. Be careful of the units of A.

4. A Chapman–Jouguet detonation is traveling at 3000 m/s. the density ratio across the wave is 12/7. The ratio of specific heats is 7/5. The pressure upstream of the wave is 1 bar and the density is 1 kg/m^3. What is the chemical heat release per unit mass?

5. Calculate the Chapman–Jouguet detonation velocity for CH_4 and stoichiometric air, assuming no dissociation. (Answer: 2305 m/s)

6. Calculate the flame temperature for methane–air flames over a range of mixture ratios. For no dissociation, calculate the mixture ratios where the flame temperature falls below the ignition temperature of 1000 K. Compare these limits with the flammability limits in Table 4.3.3a. Why are there differences?

7. Sketch the profiles of velocity, gas density, and static pressure and temperature through a deflagration and a Chapman–Jouguet detonation.

8. Prove Eq. (4.32).

9. Prove the statement above Eq. (4.7) that the reference state is transformed from 298 K to 0 K by the analysis made there.

IGNITION

5.1. INTRODUCTION

In prior chapters several methods of initiation of chemical reaction and explosion have been mentioned. These included spark, thermal, radical generation and impulsive loading initiation. In this chapter we will discuss two methods in some depth, recognizing that there are many other ways of igniting an explosive mixture. The two chosen are thermal ignition and spark ignition. Initiation by a spark is often intentional, such as in a spark ignited internal combustion engine. It is also often unintentional as, for example, with an accidental electrical discharge. In thermal ignition we usually consider an unintentional situation in which a thermal runaway occurs; heat generated by chemical reaction cannot be carried away fast enough by heat transfer to stop the chemical kinetics from accelerating.

A mixture capable of supporting an explosion or flame can always be ignited by heating it to a high enough temperature. It is only a

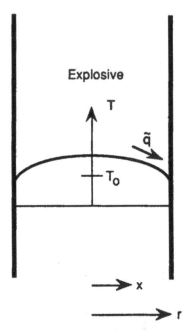

FIGURE 5.2a. Steady thermal profile between two infinite parallel plates.

question of getting the chemical kinetics going fast enough to over-whelm any heat losses. The issue in this chapter is not whether a mix-ture may be ignited, but what are the minimum requirements for an ignition source. This is the practical question from the standpoints of safety and ignition hardware and energy requirements.

5.2. THERMAL IGNITION

Consider an explosive mixture between two infinite parallel plates, as shown in Fig. 5.2a. Consider that by some means the wall temperature is held at a constant value, T_0. The only two mechanisms we consider here are heat generation by chemical reaction and internal and wall heat transfer. For a steady state to exist the processes of heat genera-tion and heat escape by heat transfer must be in balance. If the heat generation rate exceeds that of the heat loss rate, a thermal runaway will occur; that is, the mixture will explode. What we wish to show is that there is a maximum T_0 allowable and that above this temperature an explosion will occur.

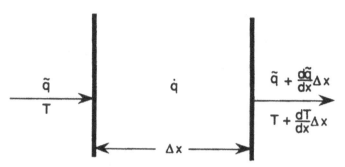

FIGURE 5.2b. Differential element with changes in temperature and heat transfer.

In Fig. 5.2b we show a differential element in x, the attendant heat transfer processes presuming a gradient in temperature exists, and we define the volumetric heat release rate as $\dot{q}$. For a steady state to exist, the net heat transfer out of the differential element must be balanced by the heat generation rate. Presuming this to be the case and using the Fourier heat conduction law

$$\frac{d}{dx}\left(-\lambda\frac{dT}{dx}\right) = \dot{q}. \tag{5.1}$$

We now define the heat release for complete consumption of the reactants as Q on a molar basis. That is Q is the heat release per mole of products. Evidently, referring back to Eq. (4.8), $Q = q \cdot W_{pr}$ since q was defined as heat release per unit mass of mixture. Originally, q was defined in terms of enthalpies of formation and here, because any explosion would take place at constant volume, energies of formation would be involved [see Eq. (2.5)]. However, the basis temperature of 0 K was used for q, and, if we presume thermally and calorically perfect gases, the formation energies and enthalpies are identical. This situation will be assumed. Assuming a global kinetics law where the rate is expressed in terms of moles product formed per unit volume per unit time

$$\dot{q} = Q A c_f^p c_o^q e^{-E/RT}$$

and placing this into Eq. (5.1) we obtain for an assumed constant coefficient of thermal conductivity

$$\frac{d^2T}{dx^2} = -\frac{Q}{\lambda} A c_f^p c_o^q e^{-E/RT}. \tag{5.2}$$

Now make some definitions that $\eta = x/r$, $\nu = T - T_0$ and $\theta = E\nu/(RT_0^2)$. We make the further approximation that the temperature at an explosion limit in the vessel is everywhere close to the wall temperature (which is usually true) so that $\nu \ll T_0$. Then the following approximation may be made:

$$e^{-E/RT} = e^{-E/R(T_0+\nu)} = e^{-E/[RT_0(1+\nu/T_0)]} \approx e^{-E/RT_0(1-\nu/T_0)}$$

$$= e^{E\nu/RT_0^2} \cdot e^{-E/RT_0} = e^{\theta} \cdot e^{-E/RT_0}.$$

Furthermore, introducing the definition

$$\delta = \left(\frac{Q}{\lambda}\right)\left(\frac{E}{RT_0^2}\right) r^2 A c_f^p c_0^q e^{-E/RT_0}, \tag{5.3}$$

Eq. (5.2) becomes, after some manipulation

$$\frac{d^2\theta}{d\eta^2} = -\delta e^{\theta}. \tag{5.4}$$

Equation (5.4) must have a time-independent solution if the situation is to be stable (non-explosive).

To Eq. (5.4) must be added boundary conditions. Since there is symmetry about $x = \eta = 0$, $(dT/dx)_0 = (d\theta/dx)_0 = 0$. Since the wall temperature is being held constant, at $\eta = 1$, $(\theta)_1 = 0$. The problem posed by Eq. (5.4) with the attendant boundary conditions forms a two boundary value problem, for which no solution is guaranteed in advance. Perhaps, only certain values of δ will allow a solution. This will indeed be the case.

We consider three cases of progressively increasing reaction rate, which is felt in the parameter δ.

Case 1. Zero reaction rate, $\delta = 0$.

Here Eq. (5.4) becomes $d^2\theta/d\eta^2 = 0$ with the solution $\theta = a_1\eta + a_2$ with a_1 and a_2 as constants of integration. The symmetry condition and the wall boundary condition are only satisfied if a_1 and a_2 are zero. This is the trivial solution, $\theta = 0$, where nothing happens.

Case 2. Finite reaction rate but $\delta \ll 1$.

In this case, θ is expected to be non-zero but small. We may therefore approximate e^{θ} in Eq. (5.4) by the first term in its Taylor series expansion about $\theta = 0$. That is $e^{\theta} = 1$. We have $d^2\theta/d\eta^2 = -\delta$. Integrating once, we have

$$\frac{d\theta}{d\eta} = a_1 - (\delta \cdot \eta).$$

The symmetry condition demands that $a_1 = 0$. Integrating a second time, we have

$$\theta = -\delta \frac{\eta^2}{2} + a_2.$$

At the wall $\theta = 0$, so $a_2 = \delta/2$. Therefore the solution is

$$\theta = \frac{\delta}{2}(1 - \eta^2).$$

Since a steady solution exists there is no thermal runaway and no explosion, but a reaction is occurring.

Case 3. δ still small but somewhat larger than in Case 2.

Here we expect larger temperature than in Case 2 so we adopt the two-term Taylor series expansion for $e^\theta = 1 + \theta$. The differential equation is now

$$\frac{d^2\theta}{d\eta^2} + \delta \cdot \theta = -\delta$$

with the solution

$$\theta = a_1 \sin(\delta^{1/2} \cdot \eta) + a_2 \cos(\delta^{1/2} \cdot \eta) - 1.$$

Differentiating once the temperature gradient is

$$\frac{d\theta}{dx} = \delta^{1/2}[a_1 \cos(\delta^{1/2} \cdot \eta) - a_2 \sin(\delta^{1/2} \cdot \eta)],$$

which can only match the symmetry condition if $a_1 = 0$ because $d\theta/d\eta = 0$ at the plane of symmetry and the cosine function is finite at $\eta = 0$. At the wall

$$\theta = 0 = a_2 \cos(\delta^{1/2}) - 1,$$

which yields $a_2 = 1/\cos \delta^{1/2}$. The solution is therefore

$$\theta = \frac{\cos(\delta^{1/2} \cdot \eta)}{\cos(\delta^{1/2})} - 1.$$

Now examine the solution for Case 3. It is physically correct only if $\theta > 0$ everywhere, since the temperature must be greater than the wall temperature in the slab interior. For arbitrary δ this solution will become unbounded for $\cos\delta^{1/2} = 0$ or at $\delta^{1/2} = \pi/2$. No steady solution is possible for δ greater than this critical value and an explosion occurs. Actually, the solution fails for lower values of δ because the $e^\theta = 1 + \theta$ approximation fails for large θ. If we had done this problem exactly (no closed form exact solution exists, but the problem can be done numerically) the critical value of δ is $\delta_{crit} = 0.88$. If the problem had been done for a spherical vessel with r as the sphere radius $\delta_{crit} = 3.32$.

We now return to the defining equation for δ, Eq. (5.3), and consider the parameter evaluated at its critical value. For gas concentrations we have $c \propto p$ and $n = p + q$. Consequently, for a given chemical system the dominant dependencies at the critical conditions for ignition are that

$$\frac{r_{crit}^2 p^n e^{-E/RT_0}}{T_0} < \text{constant}$$

to prevent ignition. The exponential in temperature usually dominates insofar as the temperature dependence of the result is concerned. The result says that at high enough pressure and temperature or with a large enough vessel (which minimizes heat escape) an explosion will occur. For solids or liquids the concentrations are independent of pressure, so the pressure dependence disappears.

The existence of a critical explosive size or characteristic physical dimension for ignition occurs repeatedly in combustion. We have seen it before in laminar flame quenching. There also exists a critical diameter for detonation. We will see it again in flameholding studies. It is also crucial for spark ignition, as we will now see.

Example 5.1.
For a given perfect gas reactant mixture the global activation energy is 120 kJ/mol. It is known that a thermal explosion will occur in a spherical vessel of radius 50 cm if the wall temperature is at 700 K. What percent decrease in critical radius will occur if the wall temperature is raised to 800 K?

Solution.
Since the dominant temperature dependence in the kinetics comes from the exponential, we assume from Eq. (5.5) that

$$r^2 \propto \exp(E/(RT_0))$$

or

$$2\ln(r) = E/RT_0 + \text{constant}$$

Differentiating, we have that

$$\frac{dr}{r} = \frac{-EdT_0}{2RT_0^2} = \frac{-(120 \times 10^3)(100)}{2 \times 8.314 \times 700^2} = -0.147.$$

Roughly, then, the radius decrease is 15%.

TABLE 5.2a
Ignition temperatures at 1 bar for stoichiometric mixtures[1]

Fuel	Oxidizer	Temperature (K)
Acetylene	Oxygen	569
	Air	578
Hydrogen	Oxygen	833
	Air	845
Methane	Oxygen	829
	Air	810
Propane	Oxygen	741
	Air	766

[1]B. P. Mullins in Chapter 11, Agardograph No. 4. London: Butterworth, 1955.

In thermal ignition problems the chemical kinetics are crucial. As has been seen in the case of the hydrogen–oxygen explosion in Chapter 3, above a certain temperature it is always possible to get the reaction going fast enough for an explosion to occur, regardless of heat transfer at walls. On the other hand, regardless of the size of the containment vessel, there are some explosives that are in metastable equilibrium below a certain temperature, which has been called the ignition temperature. These temperatures are basically the temperatures below which even a vessel of infinite size will not support an explosion. These temperatures are maximum safe temperatures, for supporting laboratory work with chemicals, for example. A listing of the autoignition temperature at 1 bar pressure is given for some chemical systems in Table 5.2a.

5.3. SPARK IGNITION

An electrical discharge in a gas across an electrode gap is called a spark. A schematic of the situation is shown in Fig. 5.3a. The actual

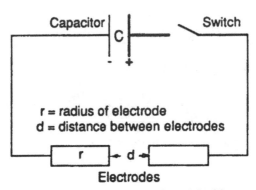

FIGURE 5.3a. Schematic of spark ignition.

physics and chemistry that take place are rather complex, involving radical generation, heating of the medium and generation of an electrically conducting gas called a *plasma*. The simplest way to think about the spark as an ignition source, however, is as a heat source. This heat is generated by the electrical resistance of the explosive mixture as the current passes through it.

Viewing Fig. 5.3a, the spark may be considered to heat a column of fluid of length d. The heat deposited, Q, may be calculated for the transient event as

$$Q = C(v_1^2 - v_2^2)/2$$

where C is the capacitance of the source, v_1 is the initial voltage and v_2 is the voltage at the end of the discharge. When d is large, a fixed Q heats more fluid, but to a lower temperature. If d is too small the quenching diameter, d_p, is approached, and even if ignition could be locally achieved the flame would quench. It is not surprising, therefore, that an optimum electrode spacing occurs for a minimum ignition energy requirement.

An actual spark discharge is a three-dimensional process, so, rather than the cylindrical volume considered above, we will assume the energy is deposited in a spherical volume of radius r. At the end of the heating process

$$Q_p = Vc_p\rho(T - T_1)$$

where properties are evaluated at the final state and 1 denotes the initial state. Although it is arguable, we presume in the above that the heating process is one at roughly constant pressure, hence the Q_p. In a non-flow situation involving heat deposition and heat transfer in a transient manner the only physical variables which enter are Q, ρ, λ, c_p, T and r with x and time, t as independent variables. The only dimensional grouping of these variables with units of time is

$$\frac{r^2}{\alpha} = \frac{r^2}{(\lambda/\rho c_p)}$$

with α being called the *thermal diffusivity*. We call $r^2/\alpha = \tau_h$ the characteristic time for the heat transfer. We now consider the following physical criterion for ignition of a premixed gas: A flame will propagate if the cool down time from T to the adiabatic flame temperature, T_b, is longer than the reaction time in V and r is at least of the order of the laminar flame reaction zone thickness. That is, the critical condition for ignition is $\tau_h > \tau_r$.

Using a spherical volume of radius r, $V = 4\pi r^3/3$, and the criterion for ignition becomes

$$\frac{r^2}{\alpha} = \frac{1}{\alpha}\left[\frac{Q_p}{\rho_3^{\frac{4}{3}}\pi c_p(T - T_1)}\right]^{2/3} > \tau_r.$$

The longest of τ_h is obtained by evaluating $T = T_b$ and the same for density, which are the values toward which the variables are going during cooldown. Now recall from Chapter 4 for a laminar flame, $\tau_r = c_{p,\text{pr}}/\mathbf{R}$ and the dominant dependence for the laminar flame speed was

$$S_L \propto \frac{\alpha W_b}{\rho_b}\mathbf{R} \times f(T)$$

where $f(T)$ is a weak function of temperature and $\alpha = \lambda_b/\rho_b c_p$. Ignoring the weak function of temperature in the above and using the laminar flame speed as a measure of the reaction rate the ignition criterion becomes

$$Q_p > \frac{\alpha^3 \rho_b (4\pi/3) c_p (T_b - T_1)(\rho_b/\rho_1)^3}{S_L^3}. \tag{5.5}$$

Accepting, as usual, that the strongest temperature dependence comes through the Arrhenius factor in S_L, the dominant dependencies of Eq. (5.5) are

$$Q_{p,\text{min}} \propto \frac{\alpha^3}{S_L^3}\rho_b \propto \frac{\lambda^3}{p^2 S_L^3}. \tag{5.6}$$

Compare this result with the expression for quenching distance in Eq.
(4.23). It is not surprising that they have similar form, because spark
ignition and quenching are both based upon a competition between
heat transfer and reaction rate. In fact, there is an interesting relation
between quenching distance and minimum spark energy of the form[1]

$$Q_{p,min} = 2400d_p^{2.5} \qquad (5.7)$$

with Q in J and d_p in m.

Example 5.2.
Perform the analysis above assuming that the spark electrode spacing
is d and the electrode radius is r. Assume that a cylinder is heated to
T by the spark, instead of the spherical assumption above. What do
you conclude about the effect of the gap width?

Solution.
Here $V = \pi r^2 d$ and $\tau_h = r^2/\alpha$ or d^2/α, depending upon which is
shorter. If r is the short dimension, then

$$\frac{r^2}{\alpha} \propto \frac{Q_p}{d} > \tau_r$$

and $Q_{p,min}$ is larger for larger gap width.

As mentioned in the Introduction, there are many other methods of
intentional ignition of explosive substances, other tha a spark. These
include pyrotechnic devices which spray hot gases and/or burning metals toward the intended combustible, pyrophoric injectants (hypergolic
with air) for hot gas generation, and detonation caps. The interested
reader is referred to more advanced material on the subject to explore
the wide variety of ignition methods.

5.4. PROBLEMS

1. Using the kinetics of Table 3.3a and the global kinetics of Example
   3.2, calculate the critical sphere ignition radius for a stoichiometric
   mixture of H_2 and O_2. Use the thermal conductivity of oxygen and
   assume a pressure of 20 bar and a wall temperature of 700 K.

2. For a gasoline–air mixture burning at 10 bar in an internal-combus-
   tion engine, it is known that the quenching distance at 1 bar is

---

[1]H. F. Calcote, C. A. Gregory, Jr., C. M. Barnett, and R. B. Gilmer, *Ind. Engrg. Chem.*
44 (1952), 2656.

1 mm. What is the minimum spark ignition energy for this mixture? Correct d_p for the pressure dependence. Compare this energy to that required to heat a gas sphere with radius of a laminar flame thickness for stoichiometric methane–air. Assume this thickness to be 1 mm and the adiabatic flame temperature of Table 2.5a maybe used as a temperature estimate.

DIFFUSION FLAMES

6.1. INTRODUCTION

Most industrial combustion chambers and engines operate with their fuel and oxidizer in an initially unmixed state. For example, in diesel and jet engines a liquid fuel spray is introduced by injection into air before combustion can occur. Moreover, for a completely burned mixture exiting the combustion chamber, the fuel, air and products of combustion must interdiffuse to form a uniform mixture. The process of combustion in these situations is called a *diffusion flame* or a *nonpremixed flame*.

In some respects, the stirred reactor of Chapter 4 could be viewed as a diffusion flame, but an important distinction is made that in the stirred reactor case diffusion was fast compared with the chemical kinetics. Here, diffusion will be the slow step in the combustion process.

Two important cases will be treated in this chapter. One will introduce the mass diffusion concepts in a completely gas phase case—that of a gaseous fuel jet exhausting into a gas oxidizer. The second, and one of great practical interest, will be that of a liquid fuel droplet which evaporates, diffuses into and then burns in an oxidizing atmosphere. These cases will introduce the reader to some of the main concepts involved in diffusion flames, but are not intended as an exhaustive treatment of the field.

6.2. GASEOUS JETS

We consider the case of Fig. 6.2a where a gas fuel jet is exhausting into a quiescent oxidizing atmosphere. For purposes of example we consider that the fuel is hydrogen and the oxidizer is oxygen. If there is no ignition source and all temperatures are below the autoignition temperature, no flame will occur. A pure mixing process will take place. We presume this to be the case before dealing with the effects of combustion.

In Fig. 6.2a the fuel and oxidizer will interdiffuse in a mixing layer which grows in thickness in the downstream direction. Close to the jet exit there is a region which is unaffected by the mixing, called the *potential core*, but ultimately mixing penetrates to the center line. As one goes downstream more and more of the fuel diffuses into the oxidizer and at "infinity" there is no fuel left, because there was a finite supply in the original jet but an infinite oxidizer sink. The mass fractions of oxidizer and fuel are also shown in Fig. 6.2a at various axial stations.

As a practical matter, and as an aid in analysis, changes of quantities in the axial direction are much slower than changes in the radial direction. Mathematically, this is expressed as $\partial/\partial x \ll \partial/\partial r$. To describe the process of mass diffusion we introduce two new velocities, V_k and v_k, called the *diffusion velocity* and *absolute velocity* of species k, respectively. They are related through the usual fluid velocity by

$$\vec{v}_K = \vec{v} + \vec{V}_K. \tag{6.1}$$

We interpret the usual velocity as the mass average fluid velocity by

$$\vec{v} = \sum_{K=1}^{M} Y_k \vec{v}_K. \tag{6.2}$$

The diffusion velocity of any species is therefore the difference between the average velocity of the fluid and the absolute velocity of the species. From Eqs. (6.1) and (6.2) it follows that a property of the

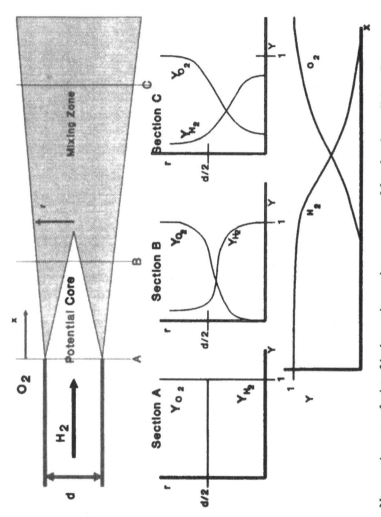

FIGURE 6.2a. Nonreacting case of a jet of hydrogen into quiescent oxygen. Mass fraction radial profile and axial profile along the flame axis.

diffusion velocities in a M-component system is

$$\sum_{K=1}^{M} Y_K \vec{V}_K = 0.$$

To describe diffusion we introduce *Fick's law* of diffusion which states

$$Y_K \vec{V}_K = -D\nabla Y_K. \tag{6.3}$$

The *diffusion coefficient*, D, is a property of the fluid just as the thermal conductivity was in the case of the Fourier law in Chapter 4. It is often true that $\lambda/\rho c_p D \approx 1$ everywhere in the field, even with variable density, λ and c_p. This dimensionless number is called the *Lewis number*.

From Eq. (6.3) and the notion that axial gradients are small, the dominant component of the gradient is radial, and we have

$$V_K = -D\frac{\partial \ln Y_K}{\partial r}. \tag{6.4}$$

In Fig. 6.2a, therefore, the hydrogen diffusion is everywhere radially outward and vice versa for the oxygen.

Now consider the case of combustion where some ignition source initiates reaction. We now make the simplifying assumption that the reaction kinetics are infinitely fast. This has as a consequence that oxygen and hydrogen cannot coexist at any field point, otherwise they would immediately react. The other consequence is that reaction must be occurring stoichiometrically so that only product (in this case H_2O) is formed. Such a process can only occur on a surface, called a *flame sheet*, where fuel and oxidizer mass fractions are zero, but the gradients are such that absolute mass flow of the fuel and oxidizer into the sheet are in stoichiometric proportions.

The situation now is as shown in Fig. 6.2b. For the stoichiometric reaction

$$H_2 + \tfrac{1}{2}O_2 \rightarrow H_2O$$

the mass flow ratio at the flame surface must be $\dot{m}_{H_2}/\dot{m}_{O_2} = \tfrac{1}{8}$. This mass flow is dominantly radial so that from Eqs. (6.1) and (6.4)

$$\rho_{H_2} v_{H_2} = -\tfrac{1}{8}\rho_{O_2} v_{O_2} = \rho_{H_2}(v + V_{H_2}) = -\tfrac{1}{8}\rho_{O_2}(v + V_{O_2})$$

$$= -\rho D\frac{\partial Y_{H_2}}{\partial r} = +\frac{1}{8}\rho D\frac{\partial Y_{O_2}}{\partial r}$$

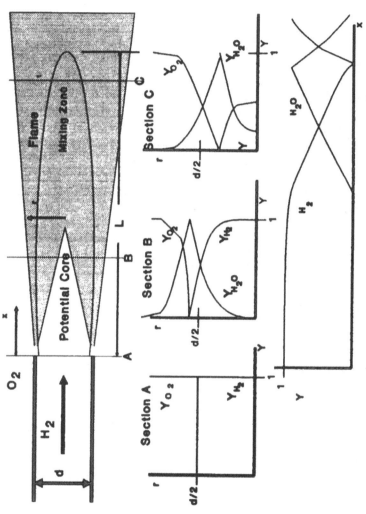

FIGURE 6.2b. Reacting case of a hydrogen jet in pure oxygen. Mass fraction profiles on radial lines and on axis.

where we have used the fact that $\rho_i = Y_i\rho$. This relation becomes an analytical boundary condition on the problem if we were to attempt a solution of the equations for species transport (which we shall not do here). Notice that it is a singular relation; the reactant mass fractions go to zero, but their gradients are finite, implying that the diffusion velocities go to infinity. This does not occur, of course, because the reaction kinetics are not truly infinitely fast. The flame sheet really has some finite thickness, but it is usually small compared with other dimensions in the flow.

This diffusion flame situation may be set up with a simple Bunsen burner in the laboratory just by cutting off the air supply. It is found in this case that the flame length is linearly proportional to the gas flow velocity. This phenomenon can be easily reasoned from the form of the diffusion relation. We argue that the end of the flame will be reached by a fluid element on the axis in a time of magnitude L_f/U. Since over most of the flame length the flame radius is of the order of d and the outward diffusion velocity of the hydrogen is given by Eq. (6.4) as $V_{H_2} \approx D/d$, a rough measure of the time for a fluid element on the axis to reach the flame is d/V_{H_2}. Equating these two times, considered as the axial convection time and the diffusion time,

$$L_f = Ud^2/D. \qquad (6.5)$$

This relation is indeed linear in U. Multiplying the numerator and denominator by ρ and noting the above observation on Lewis number, the denominator behaves like λ/c_p, which is highly insensitive to thermodynamic conditions. We conclude the dominant dependence of flame height is only upon the mass flow rate.

Example 6.1.
Hydrogen is exhausting into still air from a tube 1 cm in diameter. The hydrogen temperature is 298 K and the pressure is 1 bar. The hydrogen velocity at the tube exit is 1 m/s. The Lewis number may be assumed to be unity. Calculate the approximate flame height.

Solution.
The hydrogen density is

$$\rho = \frac{pW_{H_2}}{RT} = \frac{10^5(2)}{(8.314)(298)(1000)} = 0.081 \text{ kg/m}^3.$$

From Table 4.3a the approximate thermal conductivity is 0.174 J/(mKs) and from the JANNAF Tables the specific heat is

$$c_p = \frac{5}{2}\frac{R}{W_{H_2}} = \frac{5}{4}(8.314) = 10.35\frac{J}{gK} = 1.035 \times 10^4 \frac{J}{kgK}.$$

From Eq. (6.5)

$$L_f = \frac{(0.081)(1)(0.01)^2}{(0.174/1.035 \times 10^4)} = 0.481 \text{ m}.$$

The dependence of Eq. (6.5) only holds over a limited range of velocities. We have assumed here smooth, laminar flow. Above a certain velocity (more precisely, Reynolds number), the flow becomes rough and turbulent. Other texts treat the turbulent case.

6.3. DROPLET BURNING

Many combustion chambers operate by spraying liquid fuel into a gaseous oxidizer. Such combustors include those in diesel engines, ramjets and jet engines. Liquid propellant rocket engines can spray both the oxidizer and fuel into the combustor in the liquid phase. It usually occurs that the chemical reactions occur in the gas phase, however, after *evaporation* of the liquid phase to the vapor phase. The spray process usually involves a high pressure fuel being passed through some form of *atomizer* to form small droplets. These in turn evaporate and the gases diffuse to find oxidizer for combustion.

We first consider the process of pure evaporation without combustion. The problem we study is that of evaporation of an initially cold spherical droplet in a hot gaseous environment. The sphericity assumption is usually a good one because for the usual droplet sizes the *surface tension* is strong enough to hold the spherical shape even in the face of what might be a strong convective environment of gases flowing by the droplet.

We first consider the case of a single droplet in a quiescent gas for which the sequence is shown in Fig. 6.3a. At the first instant of injection the liquid has a uniform, cold temperature and the gas is uniformly hot, as shown in Fig. 6.3a-1. By thermal conduction, however, the drop starts to heat up. At the same time, part of the heat transfer goes toward providing the *latent heat* (enthalpy) *of evaporation*, and the gas generated starts diffusing outward. This is shown in Fig.

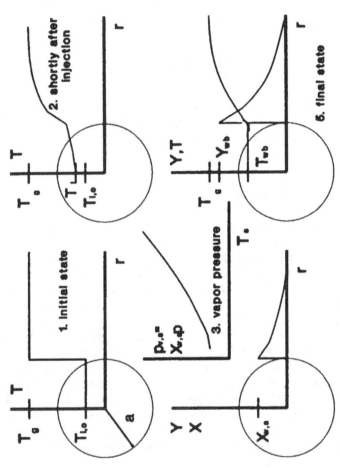

FIGURE 6.3a. Stages during the lifetime of an evaporating droplet. Temperature and mass (mole) fraction profiles and the behavior of vapor pressure.

6.3a-2. The break in slope of the temperature profile at the liquid–gas interface occurs for the same reason as in Chapter 4 for the burning solid propellant; part of the interface heat transfer goes to heating of the liquid and part goes toward providing the latent heat.

To every liquid there exists a *vapor pressure*, depending only on temperature, for which the vapor is in equilibrium with the liquid. This is shown schematically in Fig. 6.3a-3. If the vapor pressure exceeds the static pressure in the configuration chosen, the liquid is above the boiling point; it is presumed this is not the case so that the partial pressure of the vapor is below the static pressure. In any event, this is a transient process whereby the liquid is heating up and the vapor pressure (and consequently the interface vapor mass fraction) is increasing. There is ultimately reached a point, however, where the interface mass fraction is high enough to carry out the vapor by diffusion as fast as demanded by the heat transfer process providing the latent heat. At that point, the liquid temperature has reached a value called the *wet bulb temperature*, and a mass transfer, heat transfer equilibrium is achieved. There is no tendency for further liquid temperature change, and a steady state is achieved. This condition is shown in Fig. 6.3a-4.

At the equilibrium condition the heat transfer all goes toward vaporization, and the mathematical condition is

$$\dot{m}L = A_s\lambda_s\left(\frac{dT}{dr}\right)_s \qquad (6.6)$$

with $\dot{m}$ the mass vaporization rate, L the latent heat (enthalpy) and A_s the droplet surface area. The subscript s denotes gas phase properties at the interface. The interface is regressing at a rate da/dt, providing an outward gas velocity, v_g. Evidently, by continuity

$$-\rho_L\frac{da}{dt} = \rho_s v_g = \frac{\dot{m}}{A_s}. \qquad (6.7)$$

From Eqs. (6.6) and (6.7),

$$\frac{da}{dt} = -\frac{\lambda_s}{L\rho_L}\left(\frac{dT}{dr}\right)_s. \qquad (6.8)$$

Since there is only one physical length in the problem, a, and the impetus for heat transfer would logically seem to be $T_g - T_s$, we introduce dimensionless variables

$$\eta = r/a \quad \text{and} \quad \theta = (T - T_s)/(T_g - T_s).$$

Then Eq. (6.8) becomes

$$\frac{da}{dt} = -\left[\lambda_s \frac{(T_g - T_s)}{(L\rho_L a)}\right]\left(\frac{d\theta}{d\eta}\right)_s \tag{6.9}$$

and there seems to be a possibility that complete solution of the heat transfer problem would yield $(d\theta/d\eta)_s$ to be some constant.[1] If this is so then $a = a(t)$ is given by integration of Eq. (6.9) as

$$a^2 = a_0^2 - \frac{\beta t}{4} \tag{6.10}$$

with a_0 the initial drop radius and

$$\beta = \left[8\lambda_s \frac{(T_g - T_s)}{(L\rho_L)}\right]\left(\frac{d\theta}{d\eta}\right)_s. \tag{6.11}$$

That is, a major result has been achieved without solving the heat and mass transfer problem: The square of the droplet radius (or droplet diameter for that matter) decays linearly in time. The number is called the *evaporation constant*. The factor 4 arises because usually the result of Eq. (6.10) is quoted in terms of the square of the droplet diameter, not the radius. Typically for hydrocarbons evaporating in hot air β is of the order of 0.007 cm^2/sec, but the magnitude depends, of course, on the variables. The functional dependence of β on the parameters is intuitively correct, with the droplet size disappearance rate increasing for low L, low liquid density, high λ_s and high temperature differential.

The a^2-law may also be deduced from mass transfer considerations. For we have, employing Eqs. (6.1) and (6.3) and realizing that the gas

---

[1]This turns out to be true only if the vaporization rate is sufficiently low. This number is more generally a function of the dimensionless ratio $c_p(T_g - T_s)/L$, which is time independent. Consequently, the argument developed above is correct insofar as time variations are concerned, but there is, in general, a functional error in the result. This occurs because heat transfer is not the only process involved. The outward convection of gas causes a convective-diffusive balance, not one of pure heat transmission.

velocity at the interface must be wholly that of the vapor (assuming the droplet cannot be penetrated by the outside gas),

$$\dot{m} = -A_s \rho_L \frac{da}{dt} = \rho_v v_v A_s = \rho_s Y_{v,s}(v + V_v)_s A_s$$

$$= \rho_s Y_{v,s} \left(v - \frac{D}{Y_{v,s}} \frac{dY_v}{dr} \right)_s A_s$$

or

$$\rho_s v_s (1 - Y_{v,s}) = -(\rho D_s) \left(\frac{dY_v}{dr} \right)_s = -\rho_L \frac{da}{dt}(1 - Y_{v,s}). \quad (6.12)$$

The driving force for mass transfer is $Y_{v,s}$, so employ the variable $Z = Y_v/Y_{v,s}$. Then Eq. (6.12) becomes

$$a \frac{da}{dt} = \frac{(\rho D)_s Y_{v,s}}{\rho_L (1 - Y_{v,s})} \left(\frac{dZ}{d\eta} \right)_s$$

and comparing with Eq. (6.9) we obtain

$$\beta = -8 \frac{(\rho D)_s Y_{v,s}}{\rho_L (1 - Y_{v,s})} \left(\frac{dZ}{d\eta} \right)_s. \quad (6.13)$$

Equating Eq. (6.13) with Eq. (6.11), we obtain for unity Lewis number

$$\frac{c_p (T_g - T_s)}{L} = \frac{-Y_{v,s}}{(1 - Y_{v,s})} \left(\frac{dZ/d\eta}{d\theta/d\eta} \right)_s. \quad (6.14)$$

It may be shown that for unity Lewis number $(dZ/d\eta)/(d\theta/d\eta) = -1$. Equation (6.14) is nothing more or less than a relation to determine the wet bulb temperature since $Y_{v,s} = f(T_s, p)$. An approximate law, valid over limited temperature range is afforded by

$$p_v = P e^{-L/RT_s} \quad (6.15)$$

where P is a calibration constant at some temperature. Dividing Eq. (6.15) by p, we obtain

$$X_v = \frac{p_v}{p} = Y_v \frac{W_s}{W_v} = \frac{P}{p} e^{-L/RT_s}$$

which is to be used in conjunction with Eq. (6.14) to determine T_s. Exercise caution here, however. In Eq. (6.15) L is on a molar basis, but in Eq. (6.14) it is on a mass basis.

Example 6.2.
A water droplet is vaporizing in 1000 K air at a static pressure of 1 bar. The latent heat of vaporization is 1100 Btu/lb. We know the vapor pressure of water at 373 K is 1 bar (the boiling point). Calculate the wet bulb temperature. For simplicity, assume the molecular weight of water and air are the same at 20 g/mol.

Solution.
The conversion from Btu/lb to J/mol is

$$L = 1100 \text{ (Btu/lb) } 20 \text{ (g/mol) } 1055 \text{ (J/Btu) } 2.20 \times 10^{-3} \text{ (lb/g)}$$

$$= 51.1 \text{ kJ/mol.}$$

To calibrate the vapor pressure equation

$$P = (10^5) \exp(51.1/(8.314 \times 10^{-3} \cdot 373)) = 1.43 \times 10^{12} \text{ Nt/m}^2.$$

Under the stated approximation

$$Y_{v,s} = X_{v,s} = [1.43 \times 10^{12} \exp(-6146/T_s)]/10^5.$$

Placing this into Eq. (6.14) with c_p assumed to be 1.2 J/gK, we obtain,

$$20\frac{1.2(1000 - T_s)}{51,000} = \frac{Y_{v,s}}{1 - Y_{v,s}} = f(T_s).$$

This is solved by trial and error.

$$T_s = 343 \text{ K.}$$

The number $c_p(T_g - T_s)/L$ is usually given a name—*the Spalding transfer number*. It is the ratio of the enthalpy transport to the energy required for vaporization. The above has been oversimplified and is really only accurate for low value of B. Exact analysis shows the evaporation constant to be given by

$$\beta = 8\lambda(T_g - T_s)\ln(1 + B)/(\rho_L L c_p). \tag{6.16}$$

The results of Eq. (6.16) become equivalent to the results of Eq. (6.11) and (6.13) in the limit of low B where $\ln(1 + B)$ is approximately equal to B. For the details the reader is referred to more advanced texts.

We now consider the more complex case of a burning droplet. The extension is guided by the results for jets, above. Again, we consider the case of infinitely fast kinetics, so that the fuel and the oxidizer cannot coexist and a flame sheet must be formed. The situation for a pure fuel burning in a pure oxidizer must be as shown in Fig.6.3b. The surrounding flame at high temperature is now the driving force for heat transport towards the drop, and the outward mass transport is driven by the fuel vapor mass fraction gradient between the droplet surface and the flame. Exact analysis shows the a^2 law to still hold but the transfer number contains some details of the chemical reaction. Specifically, for a pure fuel burning in a pure oxidizer the transfer number is

$$B = \frac{c_p(T_g - T_s) + jh}{L} \qquad (6.17)$$

where h is the enthalpy of the stoichiometric reaction per unit fuel mass and j is the stoichiometric mass ratio of fuel to oxidizer. Typically, in combustion problems, the temperature term is unimportant compared to the heat of reaction in Eq. (6.17). This has a consequence that the wet bulb temperature need not be determined, and a reasonable approximation to T_s is that of the boiling point of the liquid at the static pressure in the combustion chamber. As a matter of fact, the boiling point approximation is usually good in very rapid pure vaporization, which occurs when the gas temperature is very high compared to the boiling temperature.

There are many complications in droplet burning and evaporation not treated by the simple calculations above. These include:

1. Real sprays usually produce droplets close enough to each other that there is interference between them. That is, the isolated droplet considerations are not valid.

2. Except in reduced gravity such as on a spacecraft natural convective flows by boyancy are introduced so that spherical symmetry is lost. This symmetry is also lost if there is relative flow between the droplets and gases in a combustion chamber, as there usually is.

3. Relative flow between the drop and gas can also induced internal circulation in the drop due to shear at the interface. The drop internal heat transfer becomes much more complex in this case.

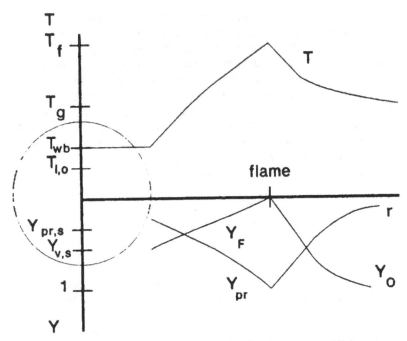

FIGURE 6.3b. Final stage of a burning droplet in a pure oxidizing atmosphere.

4. Real fuels are often mixtures of fuels, rather than being pure substances. This complicates the treatment of vaporization, because each component has its own vapor pressure properties.
5. Some combustion chambers have the liquid injected in a supercritical state, with no defined boundary between liquid and gas.
6. Some flow environments have relative velocities between the drop and gas that are so high that droplet shattering can take place. In any event, spherical symmetry may be lost for the droplet surface.
7. Much of the droplet's time may be spent in a transient state, so that the steady state considerations above do not hold.
8. When liquids are introduced by a spray, the droplets may have a whole range of sizes. Therefore, statistics of the drop size distribution have to be considered.

In spite of these problem areas the simple treatment given here gives qualitatively correct trends for liquid disappearance rate and delineates the major parameters affecting this rate. The formulas given for evaporation are probably more useful than are those for burning if one interprets the gas temperature as the combustion temperature in the combustor device. The spherical flame is almost never present in any real combustion situation.

6.4. PROBLEMS

1. Assume a fuel jet consists a mixture of 50% fuel and 50% inert gas and that it exhausts into and burns in air. Sketch the radial profiles of inert, fuel, oxygen, nitrogen and combustion products mass fractions at the location of Fig.6.1b. Also, sketch these profiles along the axis of the jet.

2. Assume a pure liquid fuel is burning in air. Sketch the profiles of fuel vapor, combustion products, nitrogen and oxygen mass fractions. Also, sketch the temperature profile.

3. It is an empirical fact that the gases in a liquid propellent rocket engine are at nearly uniform temperature throughout—the adiabatic flame temperature. Assume this temperature is 3500 K and and a fuel is vaporizing in this atmosphere. A fuel droplet, initially with a diameter of 200 mm, travels axially down the combustion chamber at a velocity of 30 m/s. The chamber pressure is 20 bar, the average gas specific heat is 1.2 J/gK, the thermal conductivity is 0.06 J/mKs and the gas molecular weight is 20 g/mol. The liquid density is 500 kg/m^3. An empirical expression for the vapor pressure is

$$p_v = 10^6 \exp[12(1 - 500/T_s)]\text{Pa}$$

where T_s is the surface temperature of the droplet in Kelvin. Assuming the droplet is at the wet bulb temperature and that the latent heat of vaporization is 100 J/g, calculate the required axial length of the combustion chamber for complete vaporization of the drop.

4. A fuel drop consisting of a mixture of two fuels may, in first approximation, be evaporated in two stages. First, the more volatile fuel is vaporized followed by the vaporization of the less volatile component.Sketch the behavior of the mass vaporization rate divided by radius and the wet bulb temperature as a function of time for the following fuel: a mixture of high vapor pressure fuel with a low latent heat and a low vapor pressure fuel with a high latent heat.

5. Two fuels of equal densities have transfer numbers of 2 and 4, respectively, but their characteristics under injection into a combustion chamber are such that their diameters are 100 μm and 200 μm, respectively. What is the ratio of their vaporization times?

FLAME
STABILIZATION

7.1. INTRODUCTION

Up until now we have assumed that there is no difficulty in establishing a stationary flame, except when flashback and blowoff were
discussed in Chapter 4 for premixed flames and when blowout was
discussed for the stirred reactor. Actually, there are rather stringent
conditions for the maintenance of a stationary flame in practical combustion devices. These conditions are of intense practical importance.
For example, the blowout of a flame in a jet engine requires some
kind of a restart procedure or the engine is of no propulsive use. The
accidental blowout of a flame in an industrial device means, at least

119

temporarily, the discharge of raw fuel and oxidizer with an attendant explosion hazard.

Blowout or blowoff of a flame is sometimes actually desired, as in the case of unintentional oil well fires. The phenomena to be discussed here are to be distinguished from intentional extinguishment of flames, such as may occur with forest or building fires. We wish to consider here the fluid mechanics and chemistry of flames with no external interaction by extinguishing agents.

7.2. AFTERBURNERS AND RAMJETS (BLUFF BODY STABILIZATION)

In order to set ideas, we first consider the stabilization of a simple laminar premixed flame on a Bunsen burner. Shown in Fig. 7.2a are, first, the velocity field of the approach gases at the mouth of the burner, and, secondly, the flame speed capability of the premixed gases as a function of radial location in the burner. The approach velocity field has a parabolic profile in laminar flow, going to zero at the wall to provide the no slip condition because of the viscosity of the real fluid. The laminar flame speed capability goes from the usual adiabatic flame speed near the center of the flow, but degrades as the wall is approached. Near the wall quenching phenomena come into play, as discussed in Chapter 4, because of heat transfer to the burner walls. Close enough to the wall (the quenching distance) there is no ability of the flame to propagate at any speed. Now, if there is any radial point at which the flow speed and the flame speed match the flame can assume a local orientation perpendicular to the approach flow. If that occurs, then the flame, locally established, can hold onto the burner at all other radial points by inclining itself to the flow, as discussed in Chapter 4. If the flow speed everywhere exceeds the flame speed, no stabilization is possible and the flame will blow off. While in principle the flame could still exist under these conditions, by appropriate inclination to the flow, there is no unique anchoring point and it may be shown that the flame is unstable in this case. The alternative case, where the flame speed exceeds the flow speed, was the case of flashback, discussed in Chapter 4. The major point here is that there is required some point in the flow that the flame can propagate against the flow perpendicular to the flow direction.

Now we consider a high speed premixed flow such as may be found in ramjet or jet engine afterburner combustion chambers. There is one major difference here compared with the laminar flows considered

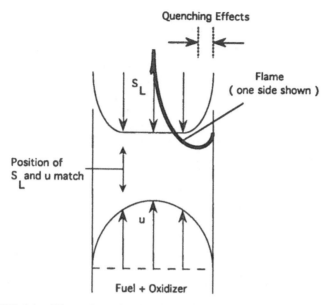

FIGURE 7.2a. Illustration of stabilization of a laminar flame on a Bunsen burner.

throughout this book to this point. The flow in most practical combustion chambers is *turbulent*. To introduce this concept, another dimensionless number of fluid mechanics must be defined, the *Reynolds number*. It measures the ratio of inertial forces to viscous forces in a flow and is given by $Re = \rho UD/\mu$. Here, the density and velocity are some characteristic values of the flow as is D, to be defined later, which is some characteristic dimension of the flow field. The *coefficient of viscosity*, μ, is a transport coefficient much like the thermal conductivity and the mass diffusivity in that it measures the transport of momentum. For most gases at normal pressures and temperatures the *Prandtl number*, $Pr = \mu c_p/\lambda$, is a number of order of unity. For example, for air Pr is approximately 0.71. Turbulence occurs at sufficiently high Reynolds numbers.

"Turbulence is an irregular (*sic.* chaotic) motion which in general makes its appearance in fluids, gaseous or liquid, when they flow past solid surfaces or even when neighboring streams of the same fluid flow past or over one another."[1] Turbulence occurs when an appro-

---

[1]Theodore von Karman, *J. Aeronaut. Sci.* **4** (1937), 131.

priately defined Reynolds number is large enough. Turbulent flows are
unsteady, but in many cases time averages of fluid quantities, such as
velocities, temperatures, etc. exist. This will be presumed to be the
case here. A primary effect of the unsteady, chaotic motion is that
transport rates of heat, mass and momentum are greater in a turbulent flow than in a laminar one. As a consequence, flame speeds in
premixed flows are faster in turbulent flows as compared with those
demonstrated in Chapter 4. Nevertheless, the flame speeds are often
much slower than flow speeds in practical devices, so that stabilization
methods must be considered.

There are many methods to stabilize premixed flames in turbulent
flows, but a common method used in ramjets and afterburners is the
use of bluff bodies inserted into the flow field. The shape is irrele-
vant as long as the body is bluff. In Fig. 7.2b a cylinder is shown as
the body; because of finite viscosity the flow cannot follow the body
smoothly in the rear half section and the flow will *separate* from the
body. This causes a recirculatory flow behind the body and is the primary requirement for flame stabilization. In fact, what is meant by
"bluff" is that the obstruction must form the recirculatory region; a
smooth flow over a streamlined body would not satisfy the necessary
conditions for stabilization. In the cold case (the upper figure in Fig.
7.2b) the length of the recirculation region is highly insensitive to the
flow conditions, as long as the flow is fully turbulent, which is usually
the case in practical applications. This length depends somewhat upon
the shape of the bluff body, but not strongly so. Typically the length of
the recirculation zone is of the order of five cylinder diameters, and
the flow speeds are substantially reduced from the freestream speed
(and are, of course, zero at the shown stagnation points).

Now consider that, by some means, combustion could be initiated.
As an empirical fact, the situation would look like the illustration at
the bottom of Fig. 7.2b. Also, the recirculation zone length is only
slightly altered by the presence of combustion. By turbulent mass
transport, some of the combustible is carried into the recirculation
zone upstream of the flame location and may be burned in this low
velocity region. By the same token, products of combustion may exit
the recirculation zone by mass transport and heat transfer can take
place from the hot recirculation zone to the oncoming fresh combustible stream, raising the temperature toward the ignition condition.
If a flame can be established near the recirculatory region, then it can
incline itself to the oncoming fresh stream away from the wake zone
much as the laminar flame was shown to do. The crucial item is in

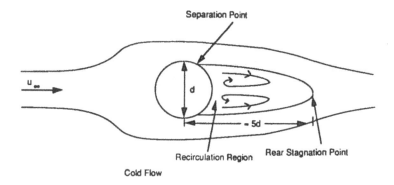

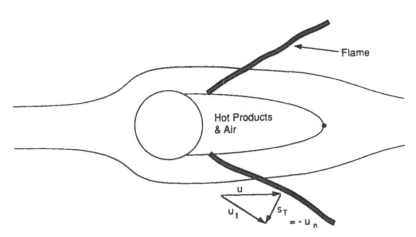

FIGURE 7.2b. Flow field and turbulent flame stabilization behind a cylindrical body.

establishing an anchor point in the low velocity region between the high speed stream and the low velocity recirculation zone.

There are several methods that have been used for analysis of this stabilization process, all of which work for scaling laws, because the situation is always a competition between residence time of the combustibles and the reaction speed. The approach used here follows the work of Zukowski and Marble.[2] With all quantities referred to their values upstream of the stabilizing device, there exists a critical blowoff

[2]E. E. Zukowski, and F. E. Marble, *Proc. Gas Dynamics Symp. Aerothermochem.*, Northwestern University, Evanston, IL, 1956, p. 205.

velocity, u_{bo} for a given dimension of the stabilizer, d, and an ignition time, τ_{ig}, of the combustible mixture. It appears reasonable that if the residence time of a fluid particle in the vicinity of the hot recirculation zone is not longer than the ignition time the flame will blowoff. If the ignition chemical kinetics is of reaction order n, then the blowoff criterion would appear to be

$$d/u_{bo} < \tau_{ig} \propto p^{-n}.$$

If, in addition, it is presumed that the ignition time and the overall reaction time are proportional (not guaranteed) then we know from prior work that $\tau_{ig} \propto \alpha/S_L^2$. We may now form an equality in dimensionless form using the blowoff condition and the ignition time condition as

$$\frac{u_{bo}d}{\alpha} = K \left(S_L \frac{d}{\alpha} \right)^2 \tag{7.1}$$

with K a proportionality constant. The reaction order has been incorporated in S_L, as usual. From an extensive correlation by Spalding[3] the constant $K = 6$ for three dimensional stabilizers such as spheres and discs. For two-dimensional objects such as cylindrical rods, K appears to be about 3. A caution here is that the flow must be such that the blowoff parameter on the left of Eq. (7.1) must be greater than 10^4. Both parameters in Eq. (7.1) are called *Peclet* numbers. They are similar to Reynolds numbers and are of the same order of magnitude as Reynolds numbers because Lewis and Prandtl numbers are of order unity for gases.

Example 7.1.
It is desired to increase the area of a ramjet combustion chamber to decrease the flow velocity by a factor of two. The flameholders are cylindrical rods and it is desired to maintain the same mixture ratio, f, in the combustion chamber. How much can the flameholder cylinder diameter be reduced in such a case?

Solution.
For the same mixture ratio, and assuming low subsonic speeds in the combustion chamber so that pressure does not change too much, all parameters in Eq. (7.1) are fixed except d and u_{bo}. If the flow speed is

---

[3]D. B. Spalding, *Some Fundamentals of Combustion.* London: Butterworths, 1955, Chap. 5.

reduced by a factor of two then the rod diameter may also be reduced by a factor of two.

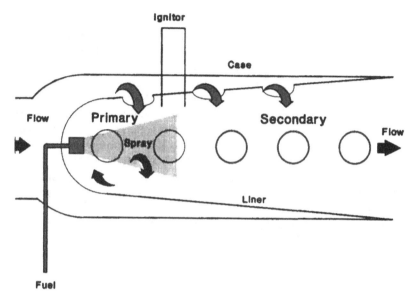

FIGURE 7.3a. Schematic of a gas turbine combustor can.

7.3. GAS TURBINE COMBUSTORS

Gas turbine combustors may have several designs but a typical form is shown in Fig. 7.3a. There are usually four major components. First, there is a case containing the high pressure air flow. Secondly, there is an atomizer or fuel injector through which high pressure liquid fuel is injected causing a fine spray of droplets.[4] Third, there is an igniter, usually a spark plug. The fourth component, the liner, has several functions and is the heart of the device concerning flame stabilization. It is a perforated device which controls the position and amount of air flow into the combustion chamber. Because the front end is usually not open to the air flow, recirculation zones must form forward of the first set of holes. In fact, because of the multiple discrete holes, there are multiple recirculation zones throughout the combustion chamber. The flow becomes highly turbulent and highly stirred. The air flow interacts strongly with the spray in the forward zone.

---

[4]It is assumed here that a liquid fuel is being used. Gas fuel may also be used.

The chamber is loosely divided into two zones, one called the primary and a downstream zone called the secondary. The hole distribution is such that the air flow into the primary, which is the section in which most of the spray is evaporated, is in nearly stoichiometric proportion to the fuel flow. Most combustion occurs here, because it is not only the hottest section of the flow field but is also the location of the fuel. The reason that the definition of the primary zone cannot be made precise is that there is a droplet size distribution to any real atomizer and some large droplets must escape the primary zone. The secondary zone lets in the rest of the air which cools down the flow to a temperature that the downstream components (the turbine) can withstand. Moreover, the mixing induced here smoothes out any hot spots or nonuniformities in temperature that may have developed in the primary. Any residual droplets also complete their combustion in this zone. The fact that relatively cool air is passing over the liner on the outside acts to cool the liner.

By this description it should be evident that the primary zone behaves much like the stirred reactor of Chapter 4. The complication here is that a spray is also involved so that chemical kinetics may not be rate controlling; spray evaporation takes time. Nevertheless, blowout of the flame can occur in much the same manner as with the stirred reactor. For a given size combustion chamber, there is a mass flow upper limit below which a flame may be sustained; otherwise, blowout occurs. With modern computational tools the design of these devices, once an empirical art, is becoming more of a science. However, it is clear that the actual flow and combustion pattern is a complex one.

7.4. PROBLEMS

1. It is desired to use ethylene fuel for a jet engine afterburner burning stoichiometrically with the turbine exhaust "air" at 3 bar pressure. The temperature entering the combustion chamber is 800 K. The flow speed is 200 m/s upstream of the flameholder section. Use the flame speed data of Fig. 4.3.2b and correct for the temperature effect assuming $S_L \propto T^4$. Assume Prandtl and Lewis numbers of unity and transport properties given elsewhere. What is the minimum two-dimensional rod flameholder diameter which will hold the flame?

2. In Problem 1, the turbulent flame speed is five times the laminar flame speed and the afterburner cross section diameter is 2 m.

How many stabilizing rods are needed to complete combustion within a length of 3 m?

3. Air at 1000 K and 20 bar enters a gas turbine combustor. Assume a constant specific heat at constant pressure of 1.2 J/gK. Assume a fuel spray of kerosene with an assumed composition of CH_2 which evaporates in a length of 12 cm, which defines the length of the primary zone. The heat of formation of the fuel is zero and it enters the combustor at 298 K. The fuel flow rate is 10 g/s. Assuming the pressure drop across the liner holes is 2% of the entrance pressure and that the velocity through the holes obeys $v = (2\Delta p/\rho)^{1/2}$, calculate the total liner hole area up to the end of the primary section. How much more hole area is needed in the secondary to drop the temperature to 1700 K?

MODERN MEASUREMENTS IN COMBUSTION

8.1. INTRODUCTION

It is presumed the student has been introduced to several intrusive methods of measurement of physical and chemical quantities in many flows. These may include methods of pressure measurement by Pitot or static probes, velocity by hot wire or film anemometry, temperature by thermocouple thermometry, etc. Measurements in combustion, however, present many unique problems. First, the environment is hostile, often at high pressure and temperature. Secondly, with chemical reactions proceeding, perturbing methods of measurements may alter the chemistry near the probe, giving bad measurements. Physical probes are also disturbing to the flow, a problem common in fluid mechanics, but often allowable in a non-reacting situation. This chapter will introduce the reader to some of the more (now) common measurement methods unique to combustion environments which address the problem areas of more usual fluid mechanics measurements.

Of interest in combustion are temperature, velocity, composition and pressure, both time average and instantaneous values. Reaction rates are also of interest and the degree of equilibrium or out of equilibrium is of paramount importance. The measurements of interest will make use of many properties of gas molecules which belong to the subject of modern physics. On the way it is hoped that the reader will understand the need for deeper understanding of the subject of the behavior of matter. This chapter will introduce the student to the subjects of statistical mechanics and quantum mechanics which should guide to further study. This chapter cannot be comprehensive. It will discuss several methods of the use of electromagnetic radiation (light) for measurements. In particular, the use of the *laser* will be introduced. It is a special light source which has outstanding properties.

The reader will recall that electromagnetic radiation may be thought of in two ways. That is, it may be considered as a wave, much as a sound wave, or as a particle, the minimum size being one *photon*, with energy $e = h\nu$, where h equals Planck's constant, 6.62×10^{-34} Js, and ν is the light frequency. The point of view taken depends upon the application. As a wave, the radiation may form *interference* patterns from multiple beams intersecting each other, and as a particle with a given energy it may excite gas molecules, as will be seen.

The laser has revolutionized combustion research in providing a non-intrusive and non-perturbing device for measurement of many quantities of interest. Laser light, which may be visible or non-visible, has the highly desirable properties or characteristics of very narrow bandwidth (single frequency or wavelength) radiation which is *coherent* (phase of the waves maintains constancy) over long path lengths. Laser beams also maintain a very low divergence angle.

Gas molecules behave according to certain rules of *quantum mechanics*, which allow only discrete energy states for the molecule. The energy may be divided into four categories. They are (a) the kinetic energy of the translation of the center of mass, (b) the rotation about the center of mass, (c) vibration about the mass center and (d) finally,

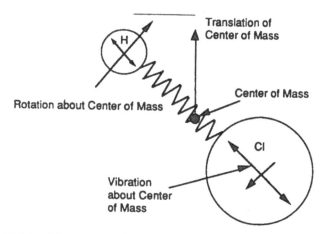

FIGURE 8.1a. Model of an HCl molecule illustrating translational, rotational and vibrational energy.

the energies of the electrons. With the exception of the *electron energies* these motions are illustrated in Fig. 8.1a. These are shown for an example molecule HCl. The kinetic energy of the center of mass translation, and its associated momentum, account for the force transmitted to a wall, for example, which is nothing more than the pressure. Vibrational and rotational energies, which are manifestations of *internal degrees of freedom* of the molecules, can be effectively used in interaction with incident light for combustion diagnostics purposes. The molecule itself may act as a *scatterer* of light, which may also be used in combustion diagnostics.

The allowability of only discrete energy levels is illustrated for vibrational energy levels in Fig. 8.1b. A *quantum number i* is assigned to each allowable level with $i = 0$ being designated the *ground state*. The levels become more and more closely spaced as i increases, and at $i = \infty$ the molecule is dissociated (infinite amplitude of vibration). Figure 8.1b will be returned to later in connection with *Raman spectroscopy* and *laser induced fluorescence* (LIF).

All molecular or atomic energy levels are quantized (discretized) and this may be used to advantage with light photons of given energies incident upon the matter. The wave behavior of light will be illustrated with *laser velocimetry* (LV) and *Rayleigh scattering*, while discrete energy levels and photon interaction will be shown in connection with Raman spectroscopy and LIF. As mentioned, this is only an introduction to some of the many modern combustion diagnostics.

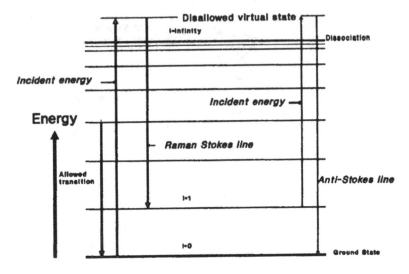

VIBRATIONAL RAMAN

FIGURE 8.1b. Illustration of vibrational energy levels and Stokes and anti-Stokes radiation emission.

8.2. LASER VELOCIMETRY

As shown in Fig. 8.2a, we consider a laser beam split into two beams (by appropriate optical components) and focused by a lens so they cross at some point. The beams have finite width, so they intersect in a finite crossing volume, called the *focal volume*. In this region an interference pattern is formed as shown in Fig. 8.2b. When waves of the same amplitude cross, they reinforce one another. This is called *constructive interference*. When a wave of positive amplitude crosses one of negative amplitude, *destructive interference* occurs. The solid lines in Fig. 8.2b for each beam indicate positive amplitude while the dashed lines indicate negative amplitude for the *wavefronts*. In the focal volume, it is seen that straight lines (planes) of destructive and constructive interference occur, causing stationary dark and light *fringes*.

If the flow into which the laser beams are focused is seeded with small particles (typically aluminum oxide in combustion) they will pass through the focal volume and scatter light, depending upon whether or not they are passing through a fringe of constructive interference. In Fig. 8.2a this scattered light is focused onto a *photomultiplier*. The time trace seen by the photomultiplier is as shown in Fig. 8.2b. Elec-

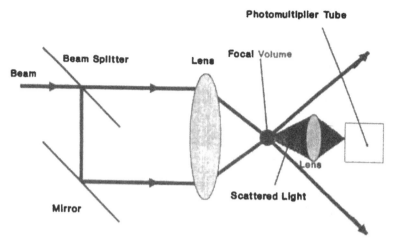

FIGURE 8.2a. Configuration of laser beams for laser velocimetry.

tronics behind the photomultiplier are capable of measuring the frequency of the alternate light and dark signals. If the particles used as seed are small enough (typically of the order of 1 μm in diameter) they will follow the flow of the gas to a good approximation. Therefore, the frequency observed in the signal measures the velocity of the flow, if the fringe spacing is known.

The geometry is as shown in Fig. 8.2b. For illustrative purposes the wavelength has been exaggerated. In reality there are many fringes in the focal volume. In fact, the electronics usually requires verification that many fringes have been crossed by a given particle for a valid measurement to have been deemed to have occurred. Shown in Fig. 8.2a is light being scattered forward of the focal volume. The actual receptor of the scattered light could be placed at almost any angle to the flow. As a practical matter, however, because of the *directionality* of the scattered light, *forward scattering* gives the best signal.

A typical laser used for velocity measurements is the Argon-ion laser, which gives beams of several wavelengths in the visible part of the light spectrum. The two strongest signals are at wavelengths of 514.5 nm in the green and 488 nm in the blue. With multiple wavelengths, by appropriate optical steering of the beams, the fringe patterns of each wavelength may be set at different angles to the flow, allowing multiple components of the velocity vector to be measured.

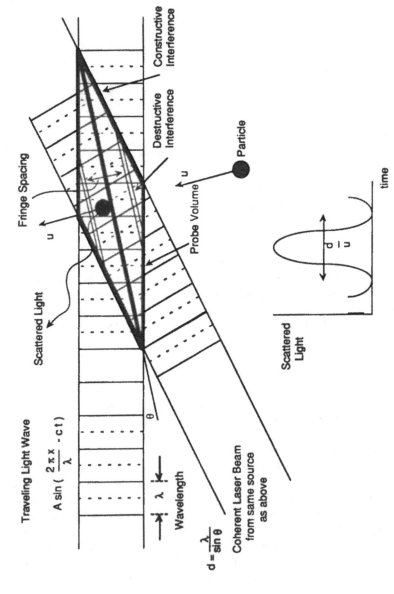

FIGURE 8.2b. Interference of two beams to form a fringe pattern for measurement of velocity.

Example 8.1.

It is desired to measure the velocity with an Argon-ion laser using the green beam in a flow with velocities of the order of 100 m/s. The half angle of the intersecting beams is 10 degrees. What is the frequency measured from the scattered light for a flow velocity of exactly 100 m/s?

Solution.

From Fig. 8.2b the fringe spacing d is

$$d = (514.5 \text{ nm})/\sin(10°) = 2950 \text{ nm}.$$

Also from that figure the frequency is

$$\nu = u/d = (100 \text{ m/s})/(2950 \times 10^{-9} \text{ m}) = 3.39 \times 10^6 \text{ Hz}.$$

Notice also in this example that for the fringe spacing of 2.95 μm, good definition of the scattered light requires seed particles smaller than this spacing.

There are several problem areas in LV, some unique to combustion. First, there is the problem of particle survivability. Second, at such small sizes, electrostatic charge may cause particle agglomeration. Third, since density of gases varies widely in combustion fields, an initial volumetric seeding of particles will become variable. That is, as the gas expands the particle density also goes down and the rate of data acquisition goes down. Fourth, with temperature variations in combustion fields, the refractive index of the gases is variable. This turns the beams as they traverse the gas so that beams which were aligned for crossing in a cold flow may become uncrossed in a hot flow. Finally, the stationary fringe pattern cannot discriminate flow direction, because only the scalar frequency is being measured. However, all of these problems may be overcome and LV is now a standard technique for velocity measurement in hot flows.

8.3. RAYLEIGH SCATTERING

Molecules, as well as particles, will scatter incident light. A relatively simple, but limited, method for measurement of either particle loading density in a gas or simply gas density in the absence of particles is through Rayleigh scattering. Simply, we measure the scattered light intensity from an incident laser beam by focusing upon an irradiated spot in the flow. This is shown in Fig. 8.3a. A pinhole is shown there

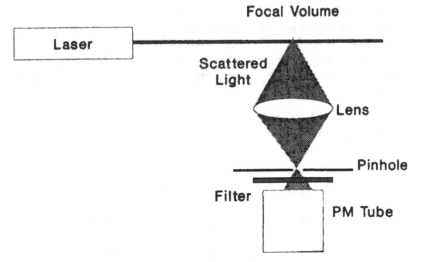

FIGURE 8.3a. Illustration of Rayleigh scattering setup.

to exclude stray light from areas other than the test volume, and a filter is shown which excludes light at frequencies other than the incident light frequency. In combustion there are often sources of radiation at many frequencies.

We will consider the case of Rayleigh scattering from gases. The first case to be investigated will be that a single species gas. To each gas there exists a *scattering cross section*, σ, such that the received scattered light, I, depends upon the incident intensity, I_0, through the relation

$$I = I_0 c \sigma \Omega. \tag{8.1}$$

The concentration in Eq. (8.1) is proportional to the molecule number density and Ω is a constant dependent solely upon the optical arrangement. Assuming the pressure is fixed, or at least known, and assuming a perfect gas, $T \propto 1/\rho \propto 1/c \propto 1/I$. If the optical system is calibrated at a known temperature condition, T_{cal}, giving an I_{cal}, then the temperature at any other condition is given by $T = T_{cal} I_{cal}/I$.

We consider next the mixing of two species at a known, fixed temperature and pressure condition. Also, consider c is known and we wish to measure X_1, the local mole fraction of species 1. Here, the scattered light is

$$I = I_0 \Omega (c_1 \sigma_1 + c_2 \sigma_2).$$

Making a calibration run at the same p and T for a pure gas, say species 1,

$$I_{cal} = I_0 \Omega c \sigma_1$$

since the overall c is the same in the two cases, through the perfect gas law. Taking the ratio of these two expressions and using the fact that $c_1 = c - c_2$

$$I/I_{cal} = 1 + (\sigma_1/\sigma_2 - 1)X_1$$

yields the mole fraction of species 1. There are tables available for the cross section values.

For flames there is usually a situation of variable T and multiple species. Nevertheless, there are still methods of using Rayleigh scattering for a T measurement. In some situations the scattering cross section is dominated by one species, because of its presence in large mole fraction. This is often the case where air is the oxidizing medium. Nitrogen dominates insofar as a species is concerned, especially in the case of fuel lean flames. Here, therefore, the cross section could be solely ascribed to the nitrogen presence, and we are back to the first method described above for a T measurement. In another case, additives could be considered which tend to make the effective σ approximately constant in going from reactants to products through a flame. This fortuitously occurs in stoichiometric methane–air flames without any additive. In a diffusion flame of a "fuel" and air, a mixture of 22% Ar and 78% H_2 has the same effective cross section as does air and the combustion products, so Rayleigh scattering may be used for T determination. It is clear that there are many "tricks" that may be used in this field.

There are several problem areas in using Rayleigh scattering in a combustion field. First, there is natural flame radiation which may interfere with the laser measurement. For example, the C_2 radical, present in hydrocarbon–oxygen flames, radiates at 515 nm, thereby competing with the 514.5 line from an argon–ion laser. Secondly, soot sometimes appears in flames; this consists of solid particles of 10 nm to 1 μm diameter. The scattering from such solids is much stronger than from gas molecules and renders Rayleigh scattering useless in such a case. Third, high T means low c and less scattering as opposed to a low T condition. Consequently, if setup for maximum signal at low T the apparatus may have low signal to noise ratio when encountering high T regions. A fourth problem, also occurring in LV, is that of beam steering by beam transmission through regions of variable refractive index. The problem is much more severe here than in LV,

however, because what is being sensed is light intensity, not just frequency. With the optics fixed, the beam jumping around in a turbulent flow, for example, takes the illuminated spot away from the focal point of the optics. Finally, many experiments have solid surfaces
which may be illuminated by stray light. This can find its way into the
receiving optics, reducing signal to noise.

8.4. RAMAN SPECTROSCOPY

Raman spectroscopy is a technique which is species specific. It requires that the discrete energy levels which gas molecules can take be
known. As such the technique can identify species concentrations directly, as well as do other things such as measure temperature. Return
to Fig 8.1b where energy levels of vibration are shown. If a molecule
is excited into a state which is not allowable, called a *virtual state*, it
will attempt to get out of the state by radiating energy to enter an
allowable state. In Raman methods an incident laser beam of high
power forces such a virtual state.

In Raman spectroscopy the lasers used are usually pulsed beams of
short duration with finite energy per pulse yielding high instantaneous
power. Typically, the laser used yields 1 J per pulse of a duration of
1 μs and with a repetition rate of 10 Hz. The laser is focused upon
a measurement volume with the intent of forcing virtual states, which
have a very low lifetime. Most molecules reside in the ground energy
state. If an energy $e = h\nu_1$ is supplied to lift the ground state to a high
virtual state and the nearly immediate decay is to the first state above
ground level ($i = 1$) the resulting energy radiated is $e = h\nu_2$. If a *spec-
trometer* is tuned to the wavelength corresponding to ν_2 the signal received is called the radiation due to the *Stokes line*. Notice that if the
molecule returns to the ground state the radiation wavelength would
be the same as the incident wavelength and this would correspond to
Rayleigh scattering. Indeed this is one advantage of the method; the
Rayleigh scattering is eliminated and the radiation received is specific
to the molecule being attacked.

If the incident radiation from the laser raises the molecule to a
virtual state from the state $i = 1$ and the decay is to the ground state
$i = 0$, the resulting signal is said to come from the *anti-Stokes* line.
The signal from this line is always substantially weaker than the Stokes
line signal, because the initial population of molecules in the $i = 1$
state is smaller than that in the $i = 0$ state.

Regardless, the basic equation for the radiated intensity is still Eq.
(8.2) with some exceptions. The cross section now depends upon not

only molecular size, but the likelihood of absorbing the incident radiation and the probability of returning to the ground state, in the case of the Stokes line. That is, probabilities are involved, not certainties. Nevertheless, the concentration of the species being attacked is directly proportional to the radiation being emitted, and this method therefore yields, after calibration, the concentration of specific species *unimpeded by the incident wavelength radiation*. Of course, the method suffers from some of the problems above, mentioned in connection with Rayleigh scattering, of, for example, beam steering and stray reflections.

Use of the Stokes and anti-Stokes lines together can yield a direct measure of the temperature. While we shall not go into detail here, the ratio of the molecular population in the $i = 1$ state to that in the $i = 0$ state is a function of temperature alone. It follows that, given an incident radiation level, the ratio of received anti-Stokes to Stokes radiation will be a function of temperature alone. While the data reduction methods are rather complex, the temperature may be deduced. The fact that the anti-Stokes line is rather weak, with attendant signal to noise ratio problems, makes this method useful only at relatively high temperatures (> 1000 K) where the $i = 1$ state has a reasonable population.

8.5. LASER INDUCED FLUORESCENCE

Returning to Fig. 8.1b, there is another way to be species specific in use of the allowable energy levels. In this case we use an exciting laser to drive the molecule between two *allowable* energy states. This requires the use of a frequency which is tuned to the desired transition. Say we drive a molecule from the $i = 0$ to $i = 4$ state using the appropriate frequency. This has produced an out-of-equilibrium situation where, on average, the molecule would like to drop into a lower state. Nevertheless, being in an allowed state, the transition takes more time than being in a virtual state, as in the case of the Raman system. If the transition, when it comes, occurs between two states at different energies than the $i = 0$ and $i = 4$ states, the radiation emitted is at a different frequency compared to that of the original radiation. This is called laser induced fluorescence, and the receiving optics are tuned to the expected radiation wavelength.

The received radiation is still proportional to the number of molecules, but the transitions are more complex and the data reduction procedures are more involved than in prior methods considered. However, LIF is especially useful when trace species are considered (such as OH) because of the species-specific radiation used and special behavior of some molecules in the fluorescence mode.

8.6. PROBLEMS

1. A Raman system delivers a pulse at a wavelength of 593 nm, and the Stokes line for a target of nitrogen molecules is at 688 nm. For each pulse 10^{-6} J is received at the photomultiplier. Calculate the number of molecules which are radiating at the Stokes line if the focusing lens intercepts 4 steradians of the focal volume.

2. Two gases are mixing at constant pressure and temperature. The ratio of the Rayleigh cross section of gas 1 to that of gas 2 is 2.0. The intensity of the received Rayleigh scattering signal in the measurement is 10^{-4} W. The calibration intensity using only gas 1 is 0.9×10^{-4} W. Calculate the mole fraction of species 1.

3. Two cylindrical laser beams intersect at a half angle of 10° forming an ellipsoid of revolution as a test volume. The beams have diameters of 1 mm. Calculate the maximum number of fringes a particle could cross traveling perpendicular to the fringe planes if the laser wavelength is 488 nm.

EPILOGUE

This book has touched upon only very few topics in the science and engineering of combustion. More advanced texts in the subject, such as those mentioned in the Preface, treat a substantially greater number of subjects than covered here and with a more rigorous base. For the reader whose interest has been piqued, the following is a list of subject areas of interest in combustion, not developed in this book:

1. The molecular theory of the thermodynamics of gases
2. The fully three-dimensional equations of change for fluids and solids
3. Imperfect gas and high pressure equation of state behavior
4. Supercritical fluid behavior
5. The molecular basis of chemical kinetics
6. Fluid–solid chemical interactions
7. Radiation heat transfer

8. Mass and heat transfer mechanisms other than Fourier and Fick's laws

9. The fluid mechanics and chemistry of pollutants such as the oxides of nitrogen and sulfer

10. Several mechanisms of flame instability

11. The combustion of solids such as coal and metals such as aluminum

12. Soot formation

13. Turbulent flames

14. Catalysis

15. The effects of gravitational buoyancy

16. Boundary layer combustion

17. Thermite reactions

18. The combustion of heterogeneous solid propellants

19. Complex hydrocarbon oxidation chemistry

20. The effects of fluid dynamic strain fields on flames

21. The calculation of transport properties

22. Fires

23. Surface tension effects in combustion

24. Supersonic combustion

25. Detonation phenomena in multiphase media

26. Vapor explosions

The above items are of intense technological interest. In further study of the subject of combustion, they will all be encountered.

APPENDIX

JANNAF THERMOCHEMICAL TABLES

The following tables are representative of data in *JANNAF Thermochemical Tables*, 3rd edition, American Institute of Physics for the National Bureau of Standards, Washington, DC, 1986. They are reproduced here, with some changes to conform to the nomenclature of this book, as an aid to solving some of the problems at the end of prior chapters. Only gas phase data are presented.

Hydrogen (H)   IDEAL GAS   W = 1.00794

Enthalpy Reference Temperature = T_r = 298.15 K   Standard State Pressure = p° = 0.1 MPa

T/K	$C_p°$	$S°$	$-[G°-H°(T_r)]/T$	$H°-H°(T_r)$	$\Delta_f H°$	$\Delta_f G°$	$\text{Log } K_f$
		— J K⁻¹mol⁻¹ —		— kJ mol⁻¹ —			
0	0.	0.	INFINITE	-8.197	216.035	216.035	INFINITE
100	20.786	92.009	133.197	-4.119	216.614	212.450	-110.972
200	20.786	106.417	116.618	-2.040	217.346	208.004	-54.325
250	20.786	111.086	115.059	-1.001	217.687	205.629	-42.964
298.15	20.786	114.716	114.716	0.	217.999	203.278	-35.613
300	20.786	114.845	114.717	0.038	218.011	203.186	-35.376
350	20.786	118.048	114.970	1.078	218.326	200.890	-29.951
400	20.786	120.825	115.532	2.117	218.637	198.150	-25.876
450	20.786	123.273	116.280	3.156	218.946	195.570	-22.701
500	20.786	125.463	117.072	4.196	219.254	192.957	-20.158
600	20.786	129.263	118.796	6.274	219.868	187.640	-16.335
700	20.786	132.457	120.524	8.363	220.479	182.220	-13.597
800	20.786	135.233	122.193	10.431	221.080	176.713	-11.538
900	20.786	137.661	123.781	12.510	221.671	171.132	-9.932
1000	20.786	139.871	125.282	14.588	222.248	165.485	-8.644
1100	20.786	141.852	126.700	16.667	222.807	159.782	-7.587
1200	20.786	143.660	128.038	18.746	223.348	154.028	-6.705
1300	20.786	145.324	129.305	20.824	223.865	148.230	-5.956
1400	20.786	146.865	130.508	22.903	224.361	142.394	-5.313
1500	20.786	148.299	131.644	24.982	224.836	136.532	-4.754
1600	20.786	149.640	132.728	27.060	225.289	130.630	-4.264
1700	20.786	150.900	133.760	29.139	225.721	124.689	-3.831
1800	20.786	152.088	134.745	31.217	226.132	118.734	-3.446
1900	20.786	153.212	135.688	33.296	226.525	112.757	-3.100
2000	20.786	154.278	136.591	35.378	226.896	106.760	-2.788
2100	20.786	155.293	137.458	37.453	227.284	100.744	-2.506
2200	20.786	156.260	138.291	39.532	227.593	94.712	-2.249
2300	20.786	157.184	139.092	41.610	227.916	88.664	-2.014
2400	20.786	158.068	139.864	43.689	228.224	82.603	-1.794
2500	20.786	158.817	140.610	45.788	228.518	76.530	-1.599
2600	20.786	159.732	141.330	47.847	228.798	70.444	-1.415
2700	20.786	160.816	142.028	49.925	229.064	64.349	-1.245
2800	20.786	161.272	142.700	52.004	229.318	58.243	-1.087
2900	20.786	162.003	143.353	54.082	229.560	52.129	-0.939
3000	20.786	162.706	143.986	56.161	229.790	46.007	-0.801
3100	20.786	163.388	144.601	58.239	230.008	39.877	-0.672
3200	20.786	164.048	145.199	60.318	230.218	33.741	-0.551
3300	20.786	164.688	145.780	62.397	230.413	27.598	-0.437
3400	20.786	165.308	146.345	64.475	230.599	21.449	-0.330
3500	20.786	165.911	146.895	66.554	230.776	15.295	-0.228
3600	20.786	166.496	147.432	68.632	230.942	9.136	-0.133
3700	20.786	167.066	147.955	70.711	231.098	2.973	-0.042
3800	20.786	167.620	148.465	72.790	231.244	-3.195	0.044
3900	20.786	168.160	148.963	74.868	231.381	-9.366	0.125
4000	20.786	168.686	149.450	76.947	231.509	-15.541	0.203
4100	20.786	169.200	149.925	79.025	231.627	-21.718	0.277
4200	20.786	169.700	150.390	81.104	231.736	-27.899	0.347
4300	20.786	170.190	150.845	83.183	231.836	-34.082	0.414
4400	20.786	170.667	151.290	85.261	231.927	-40.267	0.478
4500	20.786	171.135	151.726	87.340	232.008	-46.454	0.539
4600	20.786	171.591	152.153	89.418	232.082	-52.643	0.598
4700	20.786	172.038	152.571	91.497	232.147	-58.834	0.654
4800	20.786	172.476	152.981	93.575	232.204	-65.025	0.708
4900	20.786	172.905	153.383	95.654	232.253	-71.218	0.759
5000	20.786	173.325	153.778	97.733	232.294	-77.412	0.809
5100	20.786	173.736	154.165	99.811	232.327	-83.606	0.856
5200	20.786	174.140	154.546	101.890	232.353	-89.801	0.902
5300	20.786	174.536	154.919	103.969	232.373	-95.997	0.946
5400	20.786	174.924	155.286	106.047	232.386	-102.193	0.989
5500	20.786	175.306	155.646	108.126	232.392	-108.388	1.029
5600	20.786	175.680	156.001	110.204	232.393	-114.584	1.069
5700	20.786	176.048	156.349	112.283	232.388	-120.780	1.107
5800	20.786	176.410	156.692	114.362	232.379	-126.976	1.144
5900	20.786	176.765	157.029	116.440	232.365	-133.172	1.179
6000	20.786	177.114	157.361	118.519	232.348	-139.368	1.213

PREVIOUS: March 1977 (1 atm)   CURRENT: March 1982 (1 bar)

Hydrogen (H_2)   REFERENCE STATE - IDEAL GAS

$W = 2.01588$

Enthalpy Reference Temperature = T_r = 298.15 K   Standard State Pressure = $p° = 0.1$ MPa

T/K	$C_p°$	$S°$	$-(G°-H°(T_r))/T$	$H°-H°(T_r)$	$\Delta_f H°$	$\Delta_f G°$	Log K_f
		J K⁻¹mol⁻¹			kJ mol⁻¹		
0	0.	0.	INFINITE	-8.467	0.	0.	0.
100	28.154	100.727	155.408	-5.468	0.	0.	0.
200	27.447	119.412	133.284	-2.774	0.	0.	0.
250	28.344	125.640	131.152	-1.378	0.	0.	0.
298.15	28.836	130.680	130.680	0.	0.	0.	0.
300	28.849	130.858	130.680	0.053	0.	0.	0.
350	29.081	135.325	131.032	1.502	0.	0.	0.
400	29.181	139.216	131.817	2.959	0.	0.	0.
450	29.229	142.656	132.834	4.420	0.	0.	0.
500	29.260	145.737	133.973	5.882	0.	0.	0.
600	29.327	151.077	136.392	8.811	0.	0.	0.
700	29.441	155.606	138.822	11.749	0.	0.	0.
800	29.624	159.548	141.171	14.702	0.	0.	0.
900	29.881	163.051	143.411	17.676	0.	0.	0.
1000	30.205	166.216	145.536	20.680	0.	0.	0.
1100	30.581	169.112	147.549	23.719	0.	0.	0.
1200	30.992	171.790	149.459	26.797	0.	0.	0.
1300	31.423	174.288	151.274	29.918	0.	0.	0.
1400	31.861	176.633	153.003	33.083	0.	0.	0.
1500	32.298	178.846	154.652	36.290	0.	0.	0.
1600	32.725	180.944	156.231	39.541	0.	0.	0.
1700	33.139	182.940	157.743	42.835	0.	0.	0.
1800	33.537	184.846	159.197	46.169	0.	0.	0.
1900	33.917	186.669	160.595	49.541	0.	0.	0.
2000	34.280	188.418	161.943	52.951	0.	0.	0.
2100	34.624	190.099	163.244	56.397	0.	0.	0.
2200	34.952	191.718	164.501	59.876	0.	0.	0.
2300	35.263	193.278	165.719	63.387	0.	0.	0.
2400	35.558	194.785	166.899	66.928	0.	0.	0.
2500	35.842	196.243	168.044	70.498	0.	0.	0.
2600	36.111	197.654	169.155	74.096	0.	0.	0.
2700	36.370	199.021	170.236	77.720	0.	0.	0.
2800	36.618	200.349	171.288	81.369	0.	0.	0.
2900	36.856	201.638	172.313	85.043	0.	0.	0.
3000	37.087	202.891	173.311	88.740	0.	0.	0.
3100	37.311	204.111	174.285	92.460	0.	0.	0.
3200	37.528	205.299	175.236	96.202	0.	0.	0.
3300	37.740	206.457	176.164	99.966	0.	0.	0.
3400	37.946	207.587	177.072	103.750	0.	0.	0.
3500	38.149	208.690	177.960	107.555	0.	0.	0.
3600	38.348	209.767	178.828	111.380	0.	0.	0.
3700	38.544	210.821	179.679	115.224	0.	0.	0.
3800	38.738	211.851	180.512	119.089	0.	0.	0.
3900	38.928	212.860	181.328	122.972	0.	0.	0.
4000	39.116	213.848	182.129	126.874	0.	0.	0.
4100	39.301	214.816	182.915	130.795	0.	0.	0.
4200	39.484	215.765	183.686	134.734	0.	0.	0.
4300	39.665	216.696	184.443	138.692	0.	0.	0.
4400	39.843	217.610	185.186	142.667	0.	0.	0.
4500	40.017	218.508	185.916	146.660	0.	0.	0.
4600	40.188	219.389	186.635	150.670	0.	0.	0.
4700	40.355	220.255	187.341	154.698	0.	0.	0.
4800	40.518	221.106	188.035	158.741	0.	0.	0.
4900	40.676	221.943	188.718	162.801	0.	0.	0.
5000	40.829	222.767	189.392	166.876	0.	0.	0.
5100	40.978	223.577	190.054	170.967	0.	0.	0.
5200	41.117	224.374	190.706	175.071	0.	0.	0.
5300	41.252	225.158	191.348	179.190	0.	0.	0.
5400	41.379	225.931	191.982	183.322	0.	0.	0.
5500	41.498	226.691	192.606	187.465	0.	0.	0.
5600	41.608	227.440	193.222	191.631	0.	0.	0.
5700	41.712	228.177	193.829	195.787	0.	0.	0.
5800	41.808	228.903	194.427	199.963	0.	0.	0.
5900	41.890	229.618	195.017	204.148	0.	0.	0.
6000	41.965	230.323	195.600	208.341	0.	0.	0.

PREVIOUS: March 1961 (1 atm)   CURRENT: March 1977 (1 bar)

Hydrogen (H_2) H_2(ref)

Hydrogen Fluoride (HF)   IDEAL GAS   W = 20.006343

Enthalpy Reference Temperature = T_r = 298.15 K   Standard State Pressure = p° = 0.1 MPa

		— J K⁻¹mol⁻¹ —			— kJ mol⁻¹ —		
T/K	$C_p°$	S°	$-[G°-H°(T_r)]/T$	$H°-H°(T_r)$	$\Delta_f H°$	$\Delta_f G°$	Log K_f
0	0.	0.	INFINITE	-8.598	-272.499	-272.499	INFINITE
100	29.128	141.988	199.679	-5.772	-273.025	-273.268	142.739
200	29.128	162.148	176.445	-2.880	-273.534	-273.947	71.546
250	29.133	168.648	174.290	-1.463	-273.519	-274.304	57.313
298.15	29.136	173.780	173.780	0.	-272.846	-274.646	48.117
300	29.136	173.960	173.780	0.054	-272.848	-274.659	47.822
350	29.143	178.453	174.135	1.511	-272.609	-275.007	41.043
400	29.150	182.344	174.933	2.968	-272.696	-275.344	35.956
450	29.160	185.778	175.943	4.426	-272.602	-275.669	31.999
500	29.173	188.851	177.082	5.804	-272.934	-275.981	28.831
600	29.230	194.174	179.501	8.804	-273.205	-276.646	24.077
700	29.351	198.688	181.927	11.733	-273.522	-277.102	20.675
800	29.580	202.619	184.273	14.677	-273.858	-277.590	18.135
900	29.827	206.115	186.510	17.648	-274.205	-278.036	16.137
1000	30.160	209.275	188.631	20.644	-274.552	-278.443	14.544
1100	30.586	212.168	190.641	23.680	-274.896	-278.816	13.240
1200	30.971	214.845	192.547	26.757	-275.231	-279.157	12.151
1300	31.397	217.340	194.360	29.875	-275.556	-279.471	11.229
1400	31.833	219.683	196.086	33.036	-275.874	-279.760	10.436
1500	32.237	221.893	197.733	36.239	-276.181	-280.037	9.751
1600	32.636	223.986	199.309	39.483	-276.477	-280.273	9.150
1700	33.017	225.976	200.830	42.766	-276.763	-280.502	8.619
1800	33.376	227.873	202.270	46.085	-277.040	-280.714	8.146
1900	33.713	229.687	203.666	49.440	-277.308	-280.911	7.723
2000	34.029	231.424	205.011	52.827	-277.566	-281.093	7.341
2100	34.325	233.092	206.309	56.245	-277.815	-281.264	6.996
2200	34.601	234.696	207.563	59.692	-278.053	-281.422	6.682
2300	34.858	236.239	208.778	63.165	-278.279	-281.570	6.395
2400	35.099	237.725	209.952	66.663	-278.493	-281.709	6.131
2500	35.323	239.165	211.092	70.184	-278.694	-281.836	5.889
2600	35.533	240.556	212.198	73.727	-278.882	-281.951	5.665
2700	35.730	241.899	213.273	77.290	-279.054	-282.076	5.457
2800	35.914	243.202	214.319	80.872	-279.210	-282.185	5.264
2900	36.088	244.466	215.337	84.473	-279.349	-282.266	5.085
3000	36.251	245.692	216.329	88.090	-279.472	-282.366	4.917
3100	36.405	246.883	217.295	91.723	-279.576	-282.483	4.760
3200	36.550	248.041	218.238	95.370	-279.662	-282.575	4.613
3300	36.687	249.168	219.158	99.032	-279.730	-282.663	4.474
3400	36.818	250.265	220.057	102.706	-279.779	-282.754	4.344
3500	36.941	251.334	220.935	106.394	-279.809	-282.841	4.221
3600	37.059	252.376	221.794	110.096	-279.821	-282.937	4.106
3700	37.171	253.393	222.735	113.807	-279.814	-283.013	3.995
3800	37.279	254.386	223.487	117.530	-279.788	-283.100	3.891
3900	37.381	255.356	224.263	121.263	-279.746	-283.188	3.793
4000	37.480	256.303	225.052	125.006	-279.685	-283.277	3.699
4100	37.574	257.230	225.826	128.750	-279.607	-283.367	3.610
4200	37.665	258.137	226.584	132.531	-279.513	-283.460	3.525
4300	37.753	259.024	227.326	136.282	-279.400	-283.556	3.445
4400	37.837	259.893	228.056	140.071	-279.273	-283.654	3.367
4500	37.919	260.744	228.775	143.860	-279.130	-283.765	3.294
4600	37.996	261.578	229.479	147.655	-278.971	-283.859	3.223
4700	38.074	262.396	230.171	151.494	-278.796	-283.967	3.156
4800	38.149	263.199	230.851	155.270	-278.611	-284.079	3.091
4900	38.221	263.986	231.519	158.088	-278.410	-284.195	3.030
5000	38.291	264.759	232.176	162.914	-278.195	-284.315	2.970
5100	38.359	265.518	232.823	166.746	-277.967	-284.440	2.913
5200	38.426	266.263	233.459	170.585	-277.725	-284.569	2.859
5300	38.491	266.996	234.084	174.431	-277.471	-284.703	2.806
5400	38.555	267.716	234.701	178.284	-277.204	-284.842	2.755
5500	38.617	268.424	235.307	182.142	-276.925	-284.987	2.707
5600	38.678	269.120	235.905	186.007	-276.633	-285.135	2.660
5700	38.737	269.806	236.494	189.878	-276.329	-285.390	2.614
5800	38.796	270.480	237.074	193.754	-276.013	-285.450	2.571
5900	38.853	271.143	237.646	197.637	-275.684	-285.616	2.529
6000	38.910	271.797	238.209	201.525	-275.343	-285.787	2.488

PREVIOUS: June 1977 (1 atm)   CURRENT: June 1977 (1 bar)

Hydrogen Fluoride (HF) $F_1H_1(g)$

Carbon (C)   (REFERENCE STATE - GRAPHITE, Spectroscopic-Grade Acheson)   W = 12.011

Enthalpy Reference Temperature = T_r = 298.15 K   Standard State Pressure = $p°$ = 0.1 MPa

T/K	$C_p°$	$S°$	$-[G°-H°(T_r)]/T$	$H°-H°(T_r)$	$\Delta_f H°$	$\Delta_f G°$	Log K_f
		J K⁻¹mol⁻¹			kJ mol⁻¹		
0	0.	0.	INFINITE	-1.051	0.	0.	0.
100	1.674	0.952	10.867	-0.991	0.	0.	0.
200	5.006	3.082	6.407	-0.665	0.	0.	0.
250	6.816	4.384	5.871	-0.349	0.	0.	0.
298.15	8.517	5.740	5.740	0.	0.	0.	0.
300	8.581	5.793	5.741	0.016	0.	0.	0.
350	10.241	7.342	5.851	0.487	0.	0.	0.
400	11.817	8.713	6.117	1.039	0.	0.	0.
450	13.289	10.191	6.487	1.667	0.	0.	0.
500	14.623	11.662	6.932	2.365	0.	0.	0.
600	16.844	14.533	7.961	3.943	0.	0.	0.
700	18.537	17.263	9.097	5.716	0.	0.	0.
800	19.827	19.826	10.379	7.637	0.	0.	0.
900	20.824	22.221	11.475	9.672	0.	0.	0.
1000	21.610	24.457	12.662	11.795	0.	0.	0.
1100	22.244	26.548	13.831	13.989	0.	0.	0.
1200	22.766	28.508	14.973	16.240	0.	0.	0.
1300	23.204	30.348	16.085	18.538	0.	0.	0.
1400	23.576	32.080	17.167	20.879	0.	0.	0.
1500	23.904	33.715	18.216	23.263	0.	0.	0.
1600	24.191	35.270	19.234	25.658	0.	0.	0.
1700	24.448	36.744	20.221	28.090	0.	0.	0.
1800	24.681	38.149	21.178	30.547	0.	0.	0.
1900	24.895	39.489	22.107	33.026	0.	0.	0.
2000	25.094	40.771	23.008	35.525	0.	0.	0.
2100	25.278	42.000	23.883	38.044	0.	0.	0.
2200	25.453	43.180	24.734	40.581	0.	0.	0.
2300	25.618	44.315	25.561	43.134	0.	0.	0.
2400	25.775	45.408	26.365	45.704	0.	0.	0.
2500	25.926	46.464	27.148	48.289	0.	0.	0.
2600	26.071	47.483	27.911	50.889	0.	0.	0.
2700	26.212	48.470	28.654	53.503	0.	0.	0.
2800	26.348	49.428	29.379	56.131	0.	0.	0.
2900	26.481	50.353	30.086	58.773	0.	0.	0.
3000	26.611	51.253	30.777	61.427	0.	0.	0.
3100	26.738	52.127	31.451	64.095	0.	0.	0.
3200	26.863	52.976	32.111	66.775	0.	0.	0.
3300	26.986	53.807	32.756	69.467	0.	0.	0.
3400	27.106	54.614	33.387	72.172	0.	0.	0.
3500	27.225	55.401	34.005	74.889	0.	0.	0.
3600	27.342	56.170	34.610	77.617	0.	0.	0.
3700	27.458	56.921	35.203	80.357	0.	0.	0.
3800	27.574	57.655	35.784	83.109	0.	0.	0.
3900	27.688	58.372	36.354	85.872	.0.	0.	0.
4000	27.801	59.075	36.913	88.646	0.	0.	0.
4100	27.913	59.763	37.462	91.432	0.	0.	0.
4200	28.024	60.437	38.001	94.229	0.	0.	0.
4300	28.134	61.097	38.531	97.037	0.	0.	0.
4400	28.245	61.745	39.051	99.856	0.	0.	0.
4500	28.354	62.381	39.562	102.685	0.	0.	0.
4600	28.463	63.006	40.065	105.526	0.	0.	0.
4700	28.570	63.619	40.560	108.378	0.	0.	0.
4800	28.678	64.222	41.047	111.240	0.	0.	0.
4900	28.785	64.814	41.526	114.114	0.	0.	0.
5000	28.893	65.397	41.997	116.997	0.	0.	0.
5100	28.999	65.970	42.462	119.892	0.	0.	0.
5200	29.106	66.534	42.919	122.797	0.	0.	0.
5300	29.211	67.049	43.370	125.713	0.	0.	0.
5400	29.317	67.636	43.814	128.640	0.	0.	0.
5500	29.422	68.175	44.252	131.577	0.	0.	0.
5600	29.528	68.706	44.684	134.524	0.	0.	0.
5700	29.632	69.230	45.110	137.482	0.	0.	0.
5800	29.737	69.746	45.531	140.451	0.	0.	0.
5900	29.842	70.255	45.945	143.430	0.	0.	0.
6000	29.946	70.758	46.355	146.419	0.	0.	0.

PREVIOUS: March 1978 (1 atm)   CURRENT: March 1978 (1 bar)

Methane (CH$_4$) IDEAL GAS W = 16.04276

Enthalpy Reference Temperature = T$_r$ = 298.15 K   Standard State Pressure = p° = 0.1 MPa

T/K	C$_p$	S°	-[G°-H°(T$_r$)]/T	H°-H°(T$_r$)	Δ$_f$H°	Δ$_f$G°	Log K$_f$
		J K^{-1} mol^{-1}			kJ mol^{-1}		
0	0.	0.	INFINITE	-10.024	-66.911	-66.911	INFINITE
100	33.258	149.500	216.485	-6.698	-69.644	-64.353	33.615
200	33.473	172.577	189.418	-3.368	-72.027	-58.161	15.190
250	34.216	180.113	186.829	-1.679	-73.426	-54.536	11.385
298.15	35.639	186.251	186.251	0.	-74.873	-50.768	8.894
300	35.706	186.472	186.252	0.066	-74.929	-50.618	8.813
350	37.874	192.131	186.694	1.903	-76.461	-46.445	6.932
400	40.500	197.356	187.704	3.861	-77.969	-42.054	5.492
450	43.374	202.291	189.053	5.987	-79.422	-37.476	4.350
500	46.342	207.014	190.614	8.300	-80.802	-32.741	3.420
600	52.227	215.987	194.103	13.130	-83.308	-22.867	1.993
700	57.794	224.461	197.840	18.638	-85.452	-12.643	0.943
800	62.932	232.518	201.675	24.675	-87.238	-2.115	0.138
900	67.601	240.206	205.532	31.305	-88.692	8.616	-0.500
1000	71.795	247.549	209.370	38.170	-89.849	19.492	-1.018
1100	75.529	254.570	213.162	45.549	-90.780	30.472	-1.447
1200	78.833	261.287	216.895	53.270	-91.437	41.524	-1.807
1300	81.744	267.714	220.556	61.302	-91.945	52.626	-2.115
1400	84.306	273.868	224.146	69.606	-92.308	63.761	-2.379
1500	86.556	279.763	227.660	78.153	-92.553	74.918	-2.608
1600	88.537	285.413	231.096	86.910	-92.703	86.088	-2.810
1700	90.283	290.834	234.450	95.853	-92.780	97.265	-2.989
1800	91.824	296.039	237.728	104.960	-92.797	108.445	-3.147
1900	93.188	301.041	240.930	114.212	-92.770	119.624	-3.288
2000	94.399	305.853	244.057	123.502	-92.709	130.802	-3.416
2100	95.477	310.485	247.110	133.087	-92.634	141.975	-3.531
2200	96.438	314.949	250.093	142.684	-92.521	153.144	-3.636
2300	97.301	319.255	253.007	152.371	-92.409	164.308	-3.732
2400	98.075	323.413	255.854	162.141	-92.281	175.467	-3.819
2500	98.773	327.431	258.638	171.984	-92.174	186.622	-3.899
2600	99.401	331.317	261.350	181.893	-92.060	197.771	-3.973
2700	99.971	335.080	264.020	191.862	-91.954	208.919	-4.042
2800	100.489	338.725	266.623	201.885	-91.857	220.058	-4.105
2900	100.960	342.260	269.171	211.956	-91.773	231.196	-4.164
3000	101.389	345.690	271.664	222.076	-91.705	242.332	-4.219
3100	101.782	349.021	274.106	232.235	-91.653	253.465	-4.271
3200	102.143	352.258	276.498	242.431	-91.621	264.598	-4.319
3300	102.474	355.406	278.842	252.662	-91.609	275.730	-4.364
3400	102.778	358.470	281.139	262.925	-91.619	286.861	-4.407
3500	103.060	361.453	283.391	273.217	-91.654	297.993	-4.447
3600	103.319	364.360	285.600	283.536	-91.713	309.127	-4.485
3700	103.560	367.194	287.767	293.881	-91.796	320.262	-4.521
3800	103.783	369.959	289.894	304.246	-91.911	331.401	-4.555
3900	103.990	372.658	291.982	314.637	-92.061	342.543	-4.588
4000	104.183	375.293	294.032	325.045	-92.232	353.687	-4.619
4100	104.363	377.868	296.045	335.473	-92.422	364.836	-4.648
4200	104.531	380.385	298.023	345.918	-92.653	375.993	-4.676
4300	104.688	382.844	299.967	356.379	-92.914	387.155	-4.703
4400	104.834	385.255	301.879	366.855	-93.208	398.322	-4.729
4500	104.972	387.612	303.758	377.345	-93.533	409.497	-4.753
4600	105.101	389.921	305.608	387.849	-93.891	420.678	-4.777
4700	105.223	392.182	307.434	398.365	-94.281	431.869	-4.800
4800	105.337	394.399	309.213	408.893	-94.703	443.069	-4.822
4900	105.445	396.572	310.973	419.432	-95.156	454.277	-4.843
5000	105.546	398.703	312.707	429.982	-95.641	465.495	-4.863
5100	105.642	400.794	314.414	440.541	-96.157	476.723	-4.883
5200	105.733	402.847	316.095	451.110	-96.703	487.961	-4.902
5300	105.818	404.861	317.750	461.688	-97.278	499.210	-4.920
5400	105.899	406.840	319.382	472.274	-97.882	510.470	-4.938
5500	105.976	408.784	320.990	482.867	-98.513	521.741	-4.955
5600	106.049	410.694	322.578	493.468	-99.170	533.025	-4.972
5700	106.118	412.573	324.137	504.077	-99.852	544.330	-4.988
5800	106.184	414.418	325.678	514.692	-100.557	555.628	-5.004
5900	106.247	416.234	327.197	525.314	-101.284	566.946	-5.019
6000	106.306	418.020	328.696	535.943	-102.032	578.279	-5.034

PREVIOUS. March 1961 (1 atm)   CURRENT: March 1961 (1 bar)

Methane (CH$_4$)   C$_1$H$_4$(g)

Ethene (C₂H₄)   IDEAL GAS

W = 28.05376

Enthalpy Reference Temperature = T_r = 298.15 K   Standard State Pressure = p° = 0.1 MPa

T/K	C_p	S°	$-[G°-H°(T_r)]/T$	$H°-H°(T_r)$	$\Delta_f H°$	$\Delta_f G°$	Log K_f
	J K⁻¹ mol⁻¹			kJ mol⁻¹			
0	0.	0.	INFINITE	-10.518	60.986	60.986	INFINITE
100	33.270	180.842	363.466	-7.183	58.194	60.476	-31.589
200	35.350	203.986	222.975	-3.804	55.542	63.749	-16.649
298	38.645	213.172	230.011	-1.960	54.002	65.976	-13.785
298.15	42.886	219.330	219.330	0.	52.467	68.421	-11.987
300	43.063	219.596	219.331	0.079	52.408	68.531	-11.930
350	45.013	226.602	219.873	2.355	50.844	71.330	-10.645
400	53.048	233.343	221.138	4.882	49.354	74.360	-8.710
450	57.907	239.874	222.884	7.657	47.951	77.571	-9.004
500	62.477	246.215	224.879	10.668	46.641	80.933	-8.455
600	70.663	258.348	228.486	17.335	44.294	88.017	-7.663
700	77.714	269.783	234.408	24.763	43.300	95.467	-7.124
800	83.840	280.570	239.511	32.847	40.637	103.180	-6.737
900	89.300	290.781	244.844	41.505	39.277	111.082	-6.447
1000	93.899	300.406	249.742	50.065	38.183	119.122	-6.322
1100	98.018	309.855	254.768	60.296	37.318	127.259	-6.043
1200	101.626	318.242	259.696	70.252	36.645	135.467	-5.897
1300	104.784	326.504	264.523	80.576	36.129	143.724	-5.779
1400	107.550	334.372	269.233	91.196	35.743	152.016	-5.673
1500	109.974	341.877	273.827	102.074	35.456	160.331	-5.583
1600	112.103	349.044	278.306	113.181	35.249	168.663	-5.506
1700	113.976	355.896	282.670	124.486	35.104	177.007	-5.438
1800	115.628	362.460	286.922	135.968	35.005	185.357	-5.379
1900	117.089	368.752	291.064	147.606	34.938	193.712	-5.326
2000	118.386	374.791	295.101	159.381	34.894	202.070	-5.278
2100	119.540	380.596	299.035	171.275	34.864	210.429	-5.234
2200	120.569	386.181	302.870	183.284	34.839	218.790	-5.195
2300	121.481	391.561	306.610	195.388	34.814	227.152	-5.159
2400	122.319	396.750	310.258	207.580	34.783	235.515	-5.126
2500	123.064	401.758	313.818	219.849	34.743	243.880	-5.096
2600	123.738	406.595	317.294	232.190	34.688	252.246	-5.068
2700	134.347	411.280	320.689	244.595	34.616	260.615	-5.042
2800	134.901	415.812	324.006	257.058	34.524	268.987	-5.018
2900	125.404	420.204	327.248	269.573	34.409	277.363	-4.996
3000	125.864	424.463	330.418	282.137	34.269	285.743	-4.975
3100	126.284	428.597	333.518	294.745	34.103	294.128	-4.956
3200	126.670	432.613	336.553	307.393	33.906	302.518	-4.938
3300	127.024	436.516	339.523	320.078	33.679	310.916	-4.921
3400	127.350	440.313	342.432	332.797	33.430	319.321	-4.906
3500	127.650	444.009	345.281	345.547	33.127	327.734	-4.891
3600	127.928	447.609	348.074	358.329	32.800	336.156	-4.877
3700	128.188	451.118	350.812	371.132	32.436	344.588	-4.866
3800	128.424	454.539	353.497	383.962	32.035	353.030	-4.853
3900	128.646	457.878	356.130	396.816	31.596	361.482	-4.842
4000	128.852	461.138	358.715	409.691	31.118	369.947	-4.831
4100	129.045	464.332	361.252	422.586	30.600	378.424	-4.821
4200	129.224	467.434	363.743	435.500	30.041	386.915	-4.813
4300	129.392	470.476	366.190	448.430	29.441	395.418	-4.803
4400	129.549	473.453	368.594	461.378	28.799	403.937	-4.795
4500	129.696	476.366	370.957	474.340	28.116	412.470	-4.788
4600	129.835	479.280	373.280	487.317	27.390	421.018	-4.781
4700	129.965	482.012	375.563	500.307	26.623	429.884	-4.774
4800	130.087	484.749	377.810	513.309	25.813	438.167	-4.768
4900	130.202	487.433	380.020	526.334	24.962	446.766	-4.763
5000	130.311	490.064	382.194	539.349	24.060	455.384	-4.757
5100	130.413	492.646	384.335	552.386	23.136	464.019	-4.753
5200	130.510	495.178	386.442	565.432	22.162	473.673	-4.748
5300	130.602	497.666	388.517	578.488	21.149	481.346	-4.744
5400	130.689	500.108	390.561	591.552	20.097	490.040	-4.740
5500	130.771	502.507	392.575	604.625	19.008	498.751	-4.737
5600	130.849	504.864	394.559	617.708	17.884	507.485	-4.734
5700	130.923	507.180	396.515	630.795	16.734	516.238	-4.731
5800	130.993	509.458	398.442	643.891	15.531	525.012	-4.728
5900	131.060	511.698	400.343	656.993	14.306	533.806	-4.726
6000	131.124	513.901	402.217	670.103	13.051	542.621	-4.724

PREVIOUS: September 1965 (1 atm)   CURRENT. September 1965 (1 bar)

Ethene (C₂H₄)   C₂H₄(g)

Oxygen (O)   IDEAL GAS   W = 15.9994

Enthalpy Reference Temperature = T_r = 298.15 K   Standard State Pressure = $p°$ = 0.1 MPa

T/K	$C_p°$	$S°$	$-(G°-H°(T_r))/T$	$H°-H°(T_r)$	$\Delta_f H°$	$\Delta_f G°$	Log K_f
	J K⁻¹ mol⁻¹			kJ mol⁻¹			
0	0.	0.	INFINITE	-6.725	246.790	246.790	INFINITE
100	23.703	135.947	161.131	-4.518	247.544	242.615	-126.729
200	22.734	152.153	163.065	-2.186	248.421	237.339	-61.986
250	22.246	157.170	161.421	-1.063	248.618	234.522	-49.001
298.15	21.911	161.058	161.058	0.	249.173	231.736	-40.509
300	21.901	161.194	161.059	0.041	249.187	231.628	-40.330
350	21.657	164.551	161.324	1.129	249.537	228.673	-34.138
400	21.483	167.430	161.912	2.207	249.868	225.670	-29.469
450	21.354	169.953	162.660	3.278	250.180	222.628	-25.842
500	21.257	172.197	163.511	4.343	250.474	219.549	-22.936
600	21.124	176.060	165.291	6.462	251.013	213.312	-18.570
700	21.040	179.310	167.067	8.570	251.494	206.990	-15.446
800	20.984	182.116	168.777	10.671	251.926	200.602	-13.096
900	20.944	184.585	170.399	12.767	252.320	194.163	-11.269
1000	20.915	186.790	171.930	14.860	252.682	187.681	-9.803
1100	20.893	188.782	173.373	16.950	253.018	181.165	-8.603
1200	20.877	190.599	174.734	19.039	253.332	174.619	-7.601
1300	20.864	192.270	176.019	21.126	253.627	168.047	-6.753
1400	20.853	193.816	177.236	23.212	253.906	161.453	-6.024
1500	20.845	195.254	178.390	25.296	254.171	154.840	-5.392
1600	20.838	196.599	179.486	27.381	254.421	148.210	-4.830
1700	20.833	197.862	180.530	29.464	254.659	141.564	-4.350
1800	20.830	199.053	181.527	31.547	254.884	134.905	-3.915
1900	20.827	200.179	182.479	33.630	255.097	128.234	-3.525
2000	20.826	201.247	183.391	35.713	255.399	121.553	-3.175
2100	20.827	202.263	184.266	37.796	255.488	114.860	-2.857
2200	20.830	203.232	185.106	39.878	255.667	108.159	-2.568
2300	20.835	204.158	185.914	41.962	255.838	101.450	-2.304
2400	20.841	205.045	186.693	44.045	255.992	94.734	-2.062
2500	20.851	205.896	187.444	46.130	256.139	88.012	-1.839
2600	20.862	206.714	188.170	48.216	256.277	81.284	-1.633
2700	20.877	207.502	188.871	50.303	256.405	74.551	-1.442
2800	20.894	208.261	189.550	52.391	256.525	67.814	-1.265
2900	20.914	208.995	190.208	54.481	256.637	61.072	-1.100
3000	20.937	209.704	190.846	56.574	256.741	54.327	-0.946
3100	20.963	210.391	191.466	58.669	256.838	47.578	-0.802
3200	20.991	211.057	192.068	60.767	256.929	40.826	-0.666
3300	21.022	211.704	192.653	62.867	257.011	34.071	-0.539
3400	21.056	212.333	193.223	64.971	257.094	27.315	-0.420
3500	21.092	212.943	193.777	67.079	257.169	20.555	-0.307
3600	21.130	213.537	194.318	69.190	257.241	13.794	-0.200
3700	21.170	214.117	194.845	71.305	257.309	7.030	-0.099
3800	21.213	214.682	195.360	73.424	257.373	0.263	-0.004
3900	21.257	215.234	195.862	75.547	257.436	-6.501	0.087
4000	21.302	215.772	196.353	77.675	257.496	-13.270	0.173
4100	21.349	216.299	196.834	79.806	257.554	-20.040	0.255
4200	21.397	216.814	197.303	81.945	257.611	-26.811	0.333
4300	21.446	217.318	197.762	84.087	257.666	-33.583	0.408
4400	21.495	217.812	198.213	86.234	257.720	-40.358	0.479
4500	21.546	218.295	198.654	88.386	257.773	-47.133	0.547
4600	21.596	218.769	199.086	90.543	257.825	-53.909	0.612
4700	21.647	219.234	199.510	92.705	257.876	-60.687	0.674
4800	21.697	219.690	199.925	94.872	257.926	-67.465	0.734
4900	21.748	220.138	200.333	97.045	257.974	-74.244	0.791
5000	21.799	220.578	200.734	99.232	258.021	-81.025	0.846
5100	21.849	221.010	201.127	101.405	258.066	-87.806	0.899
5200	21.899	221.435	201.514	103.592	258.110	-94.589	0.950
5300	21.949	221.853	201.893	105.784	258.150	-101.371	0.999
5400	21.997	222.264	202.267	107.982	258.189	-108.155	1.046
5500	22.045	222.668	202.634	110.164	258.224	-114.940	1.092
5600	22.093	223.065	202.995	112.391	258.256	-121.725	1.135
5700	22.139	223.457	203.351	114.602	258.282	-128.510	1.178
5800	22.184	223.842	203.701	116.818	258.304	-135.296	1.218
5900	22.229	224.223	204.046	119.038	258.321	-142.083	1.258
6000	22.273	224.596	204.385	121.264	258.333	-148.869	1.296

PREVIOUS: March 1977 (1 atm)   CURRENT: September 1982 (1 bar)

Oxygen (O$_2$) (REFERENCE STATE - IDEAL GAS) W = 31.9988

Enthalpy Reference Temperature = T$_r$ = 298.15 K   Standard State Pressure = p° = 0.1 MPa

T/K	C$_p$	S°	$-[G°-H°(T_r)]/T$	H°$-$H°(T$_r$)	Δ$_f$H°	Δ$_f$G°	Log K$_f$
		J K^{-1} mol^{-1}			kJ mol^{-1}		
0	0.	0.	INFINITE	-8.683	0.	0.	0.
100	29.106	173.307	231.094	-5.778	0.	0.	0.
200	29.126	193.485	207.823	-2.868	0.	0.	0.
250	29.201	199.990	205.630	-1.410	0.	0.	0.
298.15	29.376	205.147	205.147	0.	0.	0.	0.
300	29.385	205.329	205.148	0.054	0.	0.	0.
350	29.694	209.880	205.506	1.531	0.	0.	0.
400	30.106	213.871	206.308	3.025	0.	0.	0.
450	30.584	217.445	207.350	4.543	0.	0.	0.
500	31.091	220.693	208.524	6.084	0.	0.	0.
600	32.090	226.451	211.044	9.244	0.	0.	0.
700	32.981	331.466	213.611	12.499	0.	0.	0.
800	33.733	335.921	216.126	15.835	0.	0.	0.
900	34.355	239.931	218.553	19.241	0.	0.	0.
1000	34.870	243.578	220.875	22.703	0.	0.	0.
1100	35.300	246.922	223.093	26.212	0.	0.	0.
1200	35.667	250.010	225.209	29.761	0.	0.	0.
1300	35.988	252.878	227.229	33.344	0.	0.	0.
1400	36.277	255.556	229.158	36.957	0.	0.	0.
1500	36.544	258.068	231.002	40.599	0.	0.	0.
1600	36.796	260.434	232.768	44.266	0.	0.	0.
1700	37.040	262.672	234.462	47.958	0.	0.	0.
1800	37.277	264.796	236.089	51.673	0.	0.	0.
1900	37.510	266.818	237.653	55.413	0.	0.	0.
2000	37.741	268.748	239.160	59.175	0.	0.	0.
2100	37.969	270.595	240.613	62.961	0.	0.	0.
2200	38.195	272.366	242.017	66.769	0.	0.	0.
2300	38.419	274.069	243.374	70.600	0.	0.	0.
2400	38.639	275.709	244.687	74.453	0.	0.	0.
2500	38.856	277.290	245.959	78.328	0.	0.	0.
2600	39.068	278.819	247.194	82.224	0.	0.	0.
2700	39.276	280.297	248.393	86.141	0.	0.	0.
2800	39.478	281.729	249.558	90.078	0.	0.	0.
2900	39.674	283.118	250.691	94.036	0.	0.	0.
3000	39.864	284.466	251.795	98.013	0.	0.	0.
3100	40.048	285.776	252.870	102.009	0.	0.	0.
3200	40.225	287.050	253.918	106.023	0.	0.	0.
3300	40.395	288.291	254.941	110.054	0.	0.	0.
3400	40.559	289.499	255.940	114.102	0.	0.	0.
3500	40.716	290.677	256.916	118.165	0.	0.	0.
3600	40.868	291.826	257.870	122.245	0.	0.	0.
3700	41.013	292.948	258.802	126.339	0.	0.	0.
3800	41.154	294.044	259.716	130.447	0.	0.	0.
3900	41.289	295.115	260.610	134.568	0.	0.	0.
4000	41.421	296.162	261.485	138.705	0.	0.	0.
4100	41.549	297.186	262.344	142.854	0.	0.	0.
4200	41.674	298.189	263.185	147.015	0.	0.	0.
4300	41.798	299.171	264.011	151.188	0.	0.	0.
4400	41.920	300.133	264.821	155.374	0.	0.	0.
4500	42.042	301.076	265.616	158.572	0.	0.	0.
4600	42.164	302.002	266.397	163.783	0.	0.	0.
4700	42.287	302.910	267.164	168.005	0.	0.	0.
4800	42.413	303.801	267.918	172.240	0.	0.	0.
4900	42.542	304.677	268.660	176.488	0.	0.	0.
5000	42.675	305.536	269.389	180.748	0.	0.	0.
5100	42.813	306.385	270.106	185.023	0.	0.	0.
5200	42.956	307.217	270.811	189.311	0.	0.	0.
5300	43.105	308.037	271.506	193.614	0.	0.	0.
5400	43.262	308.844	272.190	197.933	0.	0.	0.
5500	43.426	309.639	272.864	202.267	0.	0.	0.
5600	43.598	310.424	273.527	206.618	0.	0.	0.
5700	43.781	311.197	274.181	210.987	0.	0.	0.
5800	43.973	311.960	274.826	215.375	0.	0.	0.
5900	44.175	312.713	275.462	219.782	0.	0.	0.
6000	44.387	313.457	276.088	224.210	0.	0.	0.

PREVIOUS: March 1977 (1 atm)   CURRENT: March 1977 (1 bar)

Oxygen (O$_2$)   O$_2$(ref)

Hydroxyl (OH)   IDEAL GAS   W = 17.00734

Enthalpy Reference Temperature = T_r = 298.15 K   Standard State Pressure = p° = 0.1 MPa

T/K	$C_p°$	$S°$	$-(G°-H°(T_r))/T$	$H°-H°(T_r)$	$\Delta_f H°$	$\Delta_f G°$	Log K_f
0	0.	0.	INFINITE	-9.172	38.390	38.390	INFINITE
100	32.627	149.590	210.950	-6.139	38.471	37.214	-19.438
200	30.777	171.392	186.471	-2.976	38.832	38.803	-9.351
250	30.383	178.463	184.204	-1.480	38.930	35.033	-7.320
298.15	29.986	183.708	183.708	0.	38.987	34.277	-6.005
300	29.977	183.894	183.709	0.055	38.988	34.246	-5.963
350	29.790	188.499	184.073	1.549	39.019	33.455	-4.993
400	29.860	192.466	184.860	3.035	39.029	32.660	-4.265
450	29.867	196.054	185.921	4.515	39.020	31.864	-3.699
500	29.831	199.066	187.062	5.992	38.995	31.070	-3.246
600	29.837	204.447	189.542	8.943	38.902	29.493	-2.568
700	29.663	209.007	192.005	11.902	38.784	27.935	-2.085
800	29.917	212.983	194.384	14.880	38.598	26.399	-1.724
900	30.264	216.526	196.651	17.898	38.416	24.884	-1.444
1000	30.676	219.736	198.801	20.935	38.230	23.391	-1.222
1100	31.124	222.680	200.840	24.024	38.046	21.916	-1.041
1200	31.586	225.408	202.775	27.180	37.867	20.458	-0.891
1300	32.046	227.955	204.615	30.342	37.697	19.014	-0.764
1400	32.492	230.348	206.368	33.569	37.536	17.583	-0.656
1500	32.917	232.602	208.043	36.839	37.381	16.162	-0.563
1600	33.319	234.740	209.645	40.151	37.234	14.753	-0.482
1700	33.694	236.771	211.163	43.502	37.093	13.352	-0.410
1800	34.044	238.707	212.657	46.888	36.955	11.960	-0.347
1900	34.369	240.557	214.078	50.310	36.819	10.575	-0.291
2000	34.670	242.337	215.446	53.762	36.688	9.197	-0.240
2100	34.950	244.028	216.787	57.243	36.551	7.826	-0.195
2200	35.209	245.656	218.043	60.752	36.416	6.462	-0.153
2300	35.449	247.228	219.276	64.285	36.278	5.103	-0.116
2400	35.673	248.741	220.474	67.841	36.137	3.750	-0.082
2500	35.881	250.203	221.635	71.419	35.992	2.404	-0.050
2600	36.075	251.613	222.761	75.017	35.843	1.063	-0.021
2700	36.256	252.976	223.855	78.633	35.688	-0.271	0.005
2800	36.426	254.300	224.918	82.267	35.530	-1.600	0.030
2900	36.586	255.581	225.954	85.918	35.368	-2.924	0.053
3000	36.736	256.824	226.962	89.584	35.194	-4.241	0.074
3100	36.878	258.031	227.945	93.265	35.017	-5.552	0.094
3200	37.013	259.203	228.904	96.960	34.834	-6.858	0.112
3300	37.140	260.344	229.839	100.667	34.644	-8.158	0.129
3400	37.261	261.455	230.753	104.387	34.448	-9.452	0.145
3500	37.376	262.537	231.645	108.119	34.246	-10.741	0.160
3600	37.486	263.591	232.518	111.863	34.037	-12.023	0.174
3700	37.592	264.620	233.372	115.617	33.821	-13.300	0.188
3800	37.693	265.624	234.208	119.381	33.599	-14.570	0.200
3900	37.791	266.604	235.026	123.155	33.371	-15.834	0.212
4000	37.885	267.562	235.827	126.938	33.136	-17.093	0.223
4100	37.976	268.499	236.613	130.732	32.894	-18.346	0.234
4200	38.064	269.415	237.383	134.534	32.646	-19.593	0.244
4300	38.150	270.311	238.138	138.345	32.391	-20.833	0.253
4400	38.233	271.189	238.879	142.164	32.130	-22.068	0.262
4500	38.315	272.050	239.607	145.991	31.862	-23.297	0.270
4600	38.394	272.893	240.322	149.827	31.587	-24.520	0.278
4700	38.472	273.719	241.025	153.670	31.305	-25.737	0.286
4800	38.549	274.530	241.713	157.521	31.017	-26.947	0.293
4900	38.625	275.336	242.391	161.380	30.722	-28.152	0.300
5000	38.699	276.107	243.057	165.246	30.420	-29.350	0.307
5100	38.773	276.874	243.713	169.120	30.111	-30.542	0.313
5200	38.846	277.637	244.358	173.001	29.796	-31.729	0.319
5300	38.919	278.368	244.993	176.889	29.473	-32.908	0.324
5400	38.991	279.086	245.617	180.784	29.144	-34.083	0.330
5500	39.063	279.813	246.233	184.687	28.807	-35.251	0.335
5600	39.134	280.517	246.839	188.597	28.464	-36.412	0.340
5700	39.206	281.210	247.436	192.514	28.113	-37.568	0.344
5800	39.278	281.892	248.024	196.438	27.756	-38.718	0.349
5900	39.350	282.564	248.604	200.369	27.391	-39.860	0.353
6000	39.433	283.326	249.175	204.308	27.019	-40.997	0.357

PREVIOUS: June 1977 (1 atm)   CURRENT: June 1977 (1 bar)

Hydroxyl (OH) $H_1O_1(g)$

Water (H_2O)   IDEAL GAS $W = 18.01528$

Enthalpy Reference Temperature = T_r = 298.15 K   **Standard State Pressure = $p°$ = 0.1 MPa**

T/K	$C_p°$	$S°$	$-[G°-H°(T_r)]/T$	$H°-H°(T_r)$	$\Delta_f H°$	$\Delta_f G°$	Log K_f
		J K⁻¹ mol⁻¹			kJ mol⁻¹		
0	0.	0.	INFINITE	-9.904	-238.921	-238.921	INFINITE
100	33.299	152.388	218.534	-6.615	-240.083	-236.584	123.579
200	33.349	175.485	191.896	-3.282	-240.900	-232.766	60.792
298.15	33.590	188.834	188.834	0.	-241.826	-228.582	40.047
300	33.596	189.042	188.835	0.062	-241.844	-228.300	39.785
400	34.262	198.788	190.159	3.452	-242.846	-223.901	29.238
500	35.226	206.534	192.685	6.925	-243.826	-219.051	22.884
600	36.325	213.052	195.550	10.501	-244.756	-214.007	18.631
700	37.495	218.739	198.465	14.192	-245.632	-208.812	15.582
800	38.721	223.825	201.322	18.002	-246.443	-203.496	13.287
900	39.987	228.459	204.084	21.938	-247.185	-198.083	11.496
1000	41.268	232.738	206.738	26.000	-247.857	-192.590	10.060
1100	42.536	236.731	209.285	30.191	-248.460	-187.033	8.881
1200	43.768	240.485	211.730	34.506	-248.897	-181.425	7.897
1300	44.945	244.035	214.080	38.942	-249.473	-175.774	7.063
1400	46.054	247.407	216.341	43.493	-249.884	-170.089	6.346
1500	47.090	250.620	218.520	48.151	-250.265	-164.376	5.724
1600	48.050	253.690	220.623	52.906	-250.592	-158.639	5.179
1700	48.935	256.630	222.655	57.758	-250.881	-152.883	4.698
1800	49.749	259.451	224.621	62.693	-251.138	-147.111	4.269
1900	50.496	262.161	226.526	67.706	-251.368	-141.325	3.885
2000	51.180	264.769	228.374	72.790	-251.575	-135.528	3.540
2100	51.823	267.282	230.167	77.941	-251.762	-129.731	3.227
2200	52.408	269.706	231.909	83.153	-251.934	-123.905	2.942
2300	52.947	272.046	233.604	88.421	-252.092	-118.082	2.682
2400	53.444	274.312	235.253	93.741	-252.239	-112.252	2.443
2500	53.904	276.503	236.860	99.108	-252.379	-106.410	2.223
2600	54.329	278.625	238.425	104.520	-252.513	-100.575	2.021
2700	54.723	280.683	239.952	109.973	-252.643	-94.729	1.833
2800	55.088	282.680	241.443	115.464	-252.771	-88.875	1.658
2900	55.430	284.619	242.899	120.990	-252.897	-83.023	1.495
3000	55.748	286.504	244.321	126.549	-253.024	-77.163	1.344
3100	56.044	288.337	245.711	132.139	-253.152	-71.296	1.201
3200	56.323	290.120	247.071	137.757	-253.282	-65.430	1.068
3300	56.583	291.858	248.402	143.403	-253.416	-59.558	0.943
3400	56.828	293.550	249.705	149.073	-253.553	-53.681	0.825
3500	57.058	295.201	250.982	154.768	-253.696	-47.801	0.713
3600	57.276	296.812	252.233	160.485	-253.844	-41.916	0.608
3700	57.480	298.384	253.459	166.222	-253.997	-36.027	0.509
3800	57.675	299.919	254.661	171.980	-254.156	-30.133	0.414
3900	57.859	301.420	255.841	177.757	-254.326	-24.236	0.325
4000	58.033	302.887	256.999	183.552	-254.501	-18.334	0.239
4100	58.199	304.322	258.136	189.363	-254.684	-12.427	0.158
4200	58.357	305.726	259.252	195.191	-254.876	-6.516	0.081
4300	58.507	307.101	260.349	201.034	-255.078	-0.600	0.007
4400	58.650	308.448	261.427	206.892	-255.288	5.320	-0.063
4500	58.787	309.767	262.486	212.764	-255.508	11.245	-0.131
4600	58.918	311.061	263.528	218.650	-255.738	17.175	-0.195
4700	59.044	312.329	264.553	224.548	-255.978	23.111	-0.257
4800	59.164	313.574	265.562	230.458	-256.229	29.052	-0.316
4900	59.275	314.795	266.554	236.380	-256.491	34.998	-0.373
5000	59.390	315.993	267.531	242.313	-256.763	40.949	-0.428
5100	59.509	317.171	268.493	248.258	-257.046	46.906	-0.480
5200	59.623	318.327	269.440	254.215	-257.338	52.869	-0.531
5300	59.746	319.464	270.373	260.184	-257.639	58.838	-0.580
5400	59.864	320.582	271.293	266.164	-257.950	64.811	-0.627
5500	59.982	321.682	272.199	272.157	-258.268	70.791	-0.672
5600	60.100	322.764	273.092	278.161	-258.585	76.777	-0.716
5700	60.218	323.828	273.973	284.177	-258.930	82.768	-0.758
5800	60.335	324.877	274.841	290.204	-259.273	88.767	-0.799
5900	60.453	325.909	275.696	296.244	-259.621	94.770	-0.839
6000	60.571	326.926	276.544	302.295	-259.977	100.780	-0.877

PREVIOUS: March 1961 (1 atm)   CURRENT: March 1979 (1 bar)

Water (H_2O) $H_2O_1(g)$

Nitrogen (N)   IDEAL GAS   W = 14.0067

Enthalpy Reference Temperature = T_r = 298.15 K   Standard State Pressure = p° = 0.1 MPa

T/K	$C_p°$	$S°$	$-(G°-H°(T_r))/T$	$H°-H°(T_r)$	$\Delta_f H°$	$\Delta_f G°$	Log K_f
	J K^{-1}mol^{-1}			kJ mol^{-1}			
0	0.	0.	INFINITE	-6.197	470.820	470.820	INFINITE
100	20.786	130.593	171.780	-4.119	471.448	466.379	-243.611
200	20.786	145.001	155.201	-2.040	472.071	461.070	-120.419
250	20.786	149.638	153.642	-1.001	472.383	458.383	-95.753
298.15	20.786	153.300	153.300	0.	472.683	455.540	-79.809
300	20.786	153.429	153.300	0.038	472.694	455.434	-79.296
350	20.786	156.833	153.554	1.078	473.005	452.533	-67.537
400	20.786	159.406	154.118	2.117	473.314	449.587	-58.710
450	20.786	161.837	154.843	3.158	473.621	446.603	-51.840
500	20.786	164.047	155.855	4.196	473.923	443.584	-46.341
600	20.786	167.836	157.379	6.274	474.510	437.461	-38.084
700	20.786	171.041	159.106	8.353	475.067	431.242	-32.190
800	20.786	173.816	160.777	10.431	475.591	424.945	-27.746
900	20.786	176.204	162.364	12.510	476.081	418.584	-24.294
1000	20.786	178.454	163.866	14.589	476.540	412.171	-21.530
1100	20.786	180.436	165.284	16.667	476.970	405.713	-19.266
1200	20.786	182.244	166.623	18.746	477.374	399.217	-17.377
1300	20.786	183.908	167.899	20.824	477.758	392.688	-15.778
1400	20.786	185.448	169.089	22.903	478.118	386.131	-14.407
1500	20.786	186.882	170.228	24.982	478.462	379.548	-13.217
1600	20.786	188.224	171.311	27.060	478.791	373.943	-12.175
1700	20.786	189.484	172.344	29.139	479.107	364.318	-11.256
1800	20.787	190.672	173.329	31.218	479.411	359.674	-10.437
1900	20.788	191.796	174.272	33.296	479.705	353.014	-9.705
2000	20.790	192.863	175.175	35.375	479.990	346.339	-9.045
2100	20.793	193.877	176.042	37.454	480.266	339.650	-8.448
2200	20.797	194.844	176.874	39.534	480.536	332.947	-7.905
2300	20.804	195.769	177.676	41.614	480.799	326.233	-7.409
2400	20.813	196.655	178.446	43.695	481.057	319.507	-6.954
2500	20.826	197.504	179.194	45.777	481.311	312.770	-6.535
2600	20.843	198.322	179.914	47.860	481.561	306.024	-6.148
2700	20.864	199.109	180.610	49.945	481.809	299.268	-5.790
2800	20.891	199.868	181.285	52.033	482.054	292.503	-5.457
2900	20.924	200.601	181.938	54.124	482.299	285.728	-5.147
3000	20.963	201.311	182.572	56.218	482.543	278.946	-4.857
3100	21.010	202.000	183.188	58.317	482.789	272.155	-4.586
3200	21.064	202.667	183.786	60.420	483.036	265.357	-4.332
3300	21.126	203.317	184.368	62.530	483.286	258.550	-4.093
3400	21.197	203.948	184.935	64.646	483.540	251.738	-3.867
3500	21.277	204.564	185.487	66.769	483.799	244.915	-3.655
3600	21.365	205.164	186.025	68.903	484.064	238.086	-3.455
3700	21.463	205.751	186.550	71.043	484.335	231.248	-3.265
3800	21.569	206.325	187.063	73.194	484.614	224.405	-3.085
3900	21.685	206.887	187.564	75.357	484.903	217.554	-2.914
4000	21.809	207.437	188.054	77.532	485.201	210.695	-2.751
4100	21.941	207.977	188.534	79.719	485.510	203.829	-2.597
4200	22.082	208.508	189.003	81.920	485.830	196.955	-2.449
4300	22.231	209.029	189.463	84.136	486.164	190.073	-2.309
4400	22.388	209.542	189.913	86.367	486.510	183.183	-2.175
4500	22.551	210.047	190.355	88.614	486.871	176.285	-2.048
4600	22.722	210.544	190.788	90.877	487.247	169.378	-1.923
4700	22.899	211.035	191.214	93.158	487.638	162.465	-1.806
4800	23.081	211.519	191.632	95.457	488.046	155.542	-1.693
4900	23.269	211.997	192.043	97.775	488.471	148.610	-1.584
5000	23.461	212.469	192.447	100.111	488.912	141.670	-1.480
5100	23.658	212.935	192.844	102.467	489.372	134.721	-1.380
5200	23.858	213.397	193.235	104.843	489.849	127.762	-1.283
5300	24.061	213.853	193.619	107.238	490.345	120.794	-1.190
5400	24.268	214.305	193.998	109.655	490.860	113.817	-1.101
5500	24.474	214.752	194.371	112.092	491.394	106.829	-1.015
5600	24.683	215.195	194.739	114.550	491.947	99.832	-0.931
5700	24.892	215.633	195.102	117.028	492.519	92.825	-0.851
5800	25.102	216.068	195.460	119.526	493.110	85.808	-0.773
5900	25.312	216.499	195.813	122.049	493.720	78.780	-0.697
6000	25.521	216.926	196.161	124.590	494.349	71.742	-0.625

PREVIOUS: March 1977 (1 atm)   CURRENT: December 1982 (1 bar)

Nitrogen (N₂)   REFERENCE STATE - IDEAL GAS   W = 28.0134

Enthalpy Reference Temperature = T_r = 298.15 K   Standard State Pressure = p° = 0.1 MPa

T/K		J K⁻¹mol⁻¹			kJ mol⁻¹		
	$C_p°$	$S°$	$-[G°-H°(T_r)]/T$	$H°-H°(T_r)$	$\Delta_f H°$	$\Delta_f G°$	Log K_f
0	0.	0.	INFINITE	-8.670	0.	0.	0.
100	29.104	159.811	217.490	-5.768	0.	0.	0.
200	29.107	179.985	194.272	-2.857	0.	0.	0.
250	29.111	186.481	192.088	-1.402	0.	0.	0.
298.15	29.124	191.609	191.609	0.	0.	0.	0.
300	29.125	191.789	191.610	0.054	0.	0.	0.
350	29.165	196.281	191.964	1.511	0.	0.	0.
400	29.249	200.181	192.753	2.971	0.	0.	0.
450	29.387	203.633	193.774	4.437	0.	0.	0.
500	29.580	206.739	194.917	5.911	0.	0.	0.
600	30.110	212.176	197.353	8.894	0.	0.	0.
700	30.754	216.866	199.813	11.937	0.	0.	0.
800	31.433	221.017	202.209	15.046	0.	0.	0.
900	32.090	224.757	204.510	18.223	0.	0.	0.
1000	32.697	228.170	206.708	21.463	0.	0.	0.
1100	33.241	231.313	208.804	24.760	0.	0.	0.
1200	33.723	234.226	210.802	28.108	0.	0.	0.
1300	34.147	236.943	212.710	31.503	0.	0.	0.
1400	34.518	239.487	214.533	34.938	0.	0.	0.
1500	34.843	241.880	216.277	38.405	0.	0.	0.
1600	35.128	244.138	217.948	41.904	0.	0.	0.
1700	35.378	246.275	219.552	45.429	0.	0.	0.
1800	35.600	248.304	221.094	48.978	0.	0.	0.
1900	35.796	250.234	222.577	52.548	0.	0.	0.
2000	35.971	252.074	224.006	56.137	0.	0.	0.
2100	36.126	253.833	225.385	59.742	0.	0.	0.
2200	36.268	255.517	226.717	63.361	0.	0.	0.
2300	36.395	257.132	228.004	66.995	0.	0.	0.
2400	36.511	258.684	229.250	70.640	0.	0.	0.
2500	36.616	260.176	230.458	74.296	0.	0.	0.
2600	36.713	261.614	231.629	77.963	0.	0.	0.
2700	36.801	263.001	232.765	81.639	0.	0.	0.
2800	36.883	264.341	233.868	85.323	0.	0.	0.
2900	36.959	265.637	234.942	89.015	0.	0.	0.
3000	37.030	266.891	235.986	92.715	0.	0.	0.
3100	37.098	268.106	237.003	96.421	0.	0.	0.
3200	37.158	269.285	237.993	100.134	0.	0.	0.
3300	37.216	270.429	238.959	103.852	0.	0.	0.
3400	37.271	271.541	239.901	107.577	0.	0.	0.
3500	37.323	272.622	240.821	111.306	0.	0.	0.
3600	37.373	273.675	241.719	115.041	0.	0.	0.
3700	37.420	274.699	242.596	118.781	0.	0.	0.
3800	37.465	275.698	243.454	122.525	0.	0.	0.
3900	37.508	276.671	244.294	126.274	0.	0.	0.
4000	37.550	277.622	245.115	130.027	0.	0.	0.
4100	37.590	278.549	245.919	133.784	0.	0.	0.
4200	37.629	279.456	246.707	137.545	0.	0.	0.
4300	37.666	280.341	247.479	141.300	0.	0.	0.
4400	37.702	281.208	248.236	145.078	0.	0.	0.
4500	37.738	282.056	248.978	148.850	0.	0.	0.
4600	37.773	282.885	249.706	152.625	0.	0.	0.
4700	37.808	283.696	250.420	156.405	0.	0.	0.
4800	37.843	284.494	251.122	160.187	0.	0.	0.
4900	37.878	285.275	251.811	163.973	0.	0.	0.
5000	37.912	286.041	252.488	167.763	0.	0.	0.
5100	37.947	286.792	253.153	171.550	0.	0.	0.
5200	37.981	287.529	253.807	175.353	0.	0.	0.
5300	38.013	288.253	254.451	179.152	0.	0.	0.
5400	38.046	288.964	255.083	182.955	0.	0.	0.
5500	38.080	289.662	255.705	186.761	0.	0.	0.
5600	38.116	290.348	256.316	190.571	0.	0.	0.
5700	38.154	291.023	256.921	194.384	0.	0.	0.
5800	38.193	291.687	257.515	198.201	0.	0.	0.
5900	38.234	292.341	258.099	202.023	0.	0.	0.
6000	38.276	292.984	258.675	205.848	0.	0.	0.

PREVIOUS: March 1977 (1 atm)   CURRENT: March 1977 (1 bar)

Nitrogen (N₂)   N₂(ref)

Nitrogen Oxide (NO)   IDEAL GAS   W = 30.0061

Enthalpy Reference Temperature = T_r = 298.15 K   Standard State Pressure = $p°$ = 0.1 MPa

T/K	$C_p°$	$S°$	$-[G°-H°(T_r)]/T$	$H°-H°(T_r)$	$\Delta_f H°$	$\Delta_f G°$	Log K_f
		J K^{-1}mol^{-1}		kJ mol^{-1}			
0	0.	0.	INFINITE	-9.192	89.775	89.775	INFINITE
100	32.302	177.031	337.757	-6.073	89.991	88.944	-46.460
200	30.420	198.747	213.601	-2.951	90.202	87.800	-22.931
250	30.025	205.468	211.251	-1.441	90.256	87.193	-18.218
298.15	29.845	210.758	210.758	0.	90.291	86.600	-15.172
300	29.841	210.943	210.759	0.055	90.292	86.577	-15.074
350	29.823	215.540	211.123	1.548	90.316	85.855	-12.826
400	29.944	219.529	211.929	3.040	90.332	85.331	-11.143
450	30.175	223.088	212.974	4.543	90.343	84.705	-9.832
500	30.486	226.263	214.145	6.088	90.352	84.079	-8.784
600	31.238	231.886	216.646	9.144	90.366	82.832	-7.210
700	32.028	236.761	219.179	12.307	90.381	81.564	-6.088
800	32.767	241.087	221.652	15.548	90.398	80.303	-5.243
900	33.422	244.985	224.031	18.858	90.417	79.041	-4.587
1000	33.987	248.536	226.307	22.229	90.437	77.775	-4.063
1100	34.468	251.799	228.476	25.653	90.457	76.508	-3.633
1200	34.877	254.816	230.549	29.120	90.476	75.238	-3.275
1300	35.236	257.621	232.525	32.626	90.493	73.969	-2.972
1400	35.524	260.243	234.412	36.164	90.508	72.687	-2.712
1500	35.780	262.703	236.217	39.729	90.518	71.425	-2.487
1600	36.002	265.019	237.945	43.318	90.525	70.151	-2.290
1700	36.195	267.206	239.603	46.929	90.526	68.878	-2.116
1800	36.364	269.282	241.196	50.557	90.522	67.605	-1.962
1900	36.514	271.252	242.725	54.201	90.511	66.332	-1.824
2000	36.647	273.128	244.199	57.858	90.494	65.060	-1.699
2100	36.767	274.919	245.618	61.530	90.469	63.788	-1.587
2200	36.874	276.632	246.990	65.212	90.438	62.519	-1.484
2300	36.971	278.273	248.315	68.904	90.398	61.251	-1.391
2400	37.060	279.849	249.596	72.606	90.350	59.984	-1.306
2500	37.141	281.363	250.837	76.316	90.295	58.720	-1.227
2600	37.216	282.822	252.039	80.034	90.231	57.458	-1.154
2700	37.285	284.227	253.205	83.759	90.160	56.199	-1.087
2800	37.350	285.585	254.338	87.491	90.081	54.943	-1.035
2900	37.410	286.896	255.438	91.229	89.994	53.689	-0.987
3000	37.466	288.165	256.508	94.973	89.899	52.439	-0.913
3100	37.519	289.395	257.549	98.722	89.798	51.192	-0.863
3200	37.570	290.587	258.563	102.477	89.689	49.948	-0.815
3300	37.617	291.744	259.551	106.236	89.574	48.708	-0.771
3400	37.663	292.867	260.514	110.000	89.451	47.472	-0.729
3500	37.706	293.960	261.454	113.768	89.323	46.239	-0.690
3600	37.747	295.022	262.373	117.541	89.189	45.010	-0.653
3700	37.787	296.057	263.269	121.318	89.049	43.784	-0.618
3800	37.825	297.065	264.145	125.098	88.903	42.563	-0.585
3900	37.862	298.048	265.002	128.883	88.752	41.346	-0.554
4000	37.898	299.008	265.840	132.671	88.596	40.132	-0.524
4100	37.933	299.944	266.660	136.462	88.434	38.922	-0.496
4200	37.966	300.858	267.464	140.257	88.268	37.717	-0.469
4300	37.999	301.752	268.251	144.056	88.097	36.515	-0.444
4400	38.031	302.626	269.022	147.857	87.922	35.318	-0.419
4500	38.062	303.481	269.778	151.662	87.741	34.124	-0.396
4600	38.092	304.318	270.520	155.469	87.556	32.934	-0.374
4700	38.122	305.137	271.248	159.280	87.366	31.749	-0.353
4800	38.151	305.940	271.962	163.094	87.171	30.568	-0.333
4900	38.180	306.727	272.664	166.910	86.970	29.391	-0.313
5000	38.208	307.499	273.353	170.730	86.765	28.218	-0.295
5100	38.235	308.256	274.030	174.552	86.553	27.049	-0.277
5200	38.262	308.998	274.695	178.377	86.336	25.884	-0.260
5300	38.289	309.728	275.349	182.204	86.113	24.724	-0.244
5400	38.316	310.443	275.993	186.034	85.881	23.568	-0.228
5500	38.342	311.147	276.625	189.867	85.644	22.416	-0.213
5600	38.367	311.838	277.248	193.703	85.399	21.269	-0.198
5700	38.393	312.517	277.861	197.541	85.146	20.125	-0.184
5800	38.418	313.185	278.464	201.381	84.884	18.987	-0.171
5900	38.443	313.842	279.058	205.224	84.613	17.853	-0.158
6000	38.468	314.488	279.643	209.070	84.331	16.724	-0.146

PREVIOUS: June 1963 (1 atm)   CURRENT: June 1963 (1 bar)

Nitrogen Oxide (NO) $N_2O_2(g)$

Nitrogen Oxide (NO₂)$\quad$ IDEAL GAS

$W = 46.0055$

Enthalpy Reference Temperature = T_r = 298.15 K				Standard State Pressure = $p°$ = 0.1 MPa			
		J K⁻¹mol⁻¹		kJ mol⁻¹			
T/K	$C_p°$	$S°$	$-[G°-H°(T_r)]/T$	$H°-H°(T_r)$	$\Delta_f H°$	$\Delta_f G°$	Log K_f
0	0.	0.	INFINITE	-10.186	35.927	35.927	INFINITE
100	33.276	202.563	271.168	-6.861	34.896	39.963	-20.874
200	34.385	225.852	243.325	-3.495	33.897	45.422	-11.863
250	35.593	233.649	240.834	-1.746	33.460	48.355	-10.103
298.15	36.974	240.034	240.034	0.	33.095	51.256	-8.980
300	37.029	240.262	240.034	0.068	33.083	51.371	-8.944
350	38.583	246.086	240.491	1.958	32.766	54.445	-8.125
400	40.171	251.342	241.524	3.527	32.512	57.560	-7.517
450	41.728	256.164	242.886	5.575	32.310	60.703	-7.046
500	43.206	260.636	244.440	6.099	32.154	63.867	-6.673
600	45.834	268.755	247.830	12.555	31.959	70.230	-6.114
700	47.986	275.988	251.345	17.250	31.878	76.616	-5.717
800	49.708	282.512	254.840	22.138	31.874	83.008	-5.420
900	51.078	288.440	258.250	27.179	31.923	89.397	-5.188
1000	52.166	293.889	261.545	32.344	32.006	95.779	-5.003
1100	53.041	298.903	264.717	37.605	32.109	102.152	-4.851
1200	53.748	303.550	267.761	42.946	32.226	108.514	-4.724
1300	54.326	307.876	270.663	48.351	32.351	114.867	-4.618
1400	54.803	311.920	273.485	53.808	32.478	121.209	-4.523
1500	55.200	315.715	276.175	59.309	32.603	127.543	-4.441
1600	55.533	319.288	278.750	64.846	32.724	133.868	-4.370
1700	55.815	322.663	281.244	70.414	32.837	140.186	-4.307
1800	56.055	325.861	283.634	76.007	32.940	146.497	-4.251
1900	56.262	328.897	285.937	81.624	33.032	152.804	-4.201
2000	56.441	331.788	288.155	87.255	33.111	159.106	-4.155
2100	56.596	334.545	290.302	92.911	33.175	165.404	-4.114
2200	56.733	337.181	292.373	98.577	33.223	171.700	-4.077
2300	56.852	339.706	294.377	104.257	33.255	177.993	-4.042
2400	56.958	342.128	296.316	109.947	33.270	184.285	-4.011
2500	57.052	344.455	298.196	115.648	33.268	190.577	-3.982
2600	57.136	346.694	300.018	121.357	33.246	196.870	-3.955
2700	57.211	348.852	301.787	127.075	33.210	203.164	-3.930
2800	57.278	350.934	303.505	132.799	33.155	209.460	-3.908
2900	57.339	352.945	305.176	138.530	33.082	215.757	-3.886
3000	57.394	354.889	306.800	144.267	32.992	222.056	-3.866
3100	57.444	356.772	308.382	150.009	32.885	228.363	-3.848
3200	57.490	358.597	309.923	155.756	32.761	234.670	-3.831
3300	57.531	360.366	311.425	161.507	32.622	240.981	-3.814
3400	57.569	362.084	312.890	167.262	32.467	247.298	-3.799
3500	57.604	363.754	314.319	173.030	32.297	253.618	-3.785
3600	57.636	365.377	315.715	178.783	32.113	259.945	-3.772
3700	57.666	366.957	317.079	184.548	31.914	266.276	-3.759
3800	57.693	368.495	318.412	190.316	31.701	272.613	-3.747
3900	57.719	369.994	319.715	196.086	31.475	278.956	-3.736
4000	57.742	371.455	320.991	201.859	31.236	285.305	-3.726
4100	57.764	372.881	322.239	207.635	30.983	291.659	-3.716
4200	57.784	374.274	323.461	213.412	30.720	298.020	-3.706
4300	57.803	375.634	324.659	219.191	30.444	304.388	-3.696
4400	57.821	376.963	325.833	224.973	30.155	310.762	-3.689
4500	57.837	378.262	326.983	230.756	29.854	317.142	-3.681
4600	57.853	379.534	328.112	236.540	29.540	323.530	-3.674
4700	57.867	380.778	329.219	242.326	29.214	329.925	-3.667
4800	57.881	381.996	330.306	248.114	28.875	336.326	-3.660
4900	57.894	383.190	331.373	253.902	28.523	342.736	-3.654
5000	57.906	384.360	332.421	259.692	28.158	349.153	-3.648
5100	57.917	385.507	333.451	265.483	27.778	355.576	-3.642
5200	57.928	386.631	334.463	271.276	27.384	362.006	-3.636
5300	57.938	387.735	335.458	277.069	26.974	368.446	-3.631
5400	57.948	388.818	336.436	282.863	26.548	374.892	-3.626
5500	57.957	389.881	337.398	288.656	26.106	381.347	-3.622
5600	57.965	390.926	338.344	294.455	25.646	387.811	-3.617
5700	57.973	391.952	339.276	300.251	25.167	394.381	-3.613
5800	57.981	392.960	340.193	306.049	24.669	400.762	-3.609
5900	57.988	393.951	341.096	311.848	24.150	407.349	-3.606
6000	57.995	394.926	341.985	317.647	23.608	413.748	-3.602

PREVIOUS: September 1964 (1 atm)$\qquad$CURRENT: September 1964 (1 bar)

Nitrogen Oxide (NO₂)$\qquad\qquad\qquad\qquad$N₂O₂(g)

Carbon Monoxide (CO) IDEAL GAS W = 28.0104

Enthalpy Reference Temperature = T_r = 298.15 K   Standard State Pressure = $p°$ = 0.1 MPa

T/K	$C_p°$	$S°$	$-(G°-H°(T_r))/T$	$H°-H°(T_r)$	$\Delta_f H°$	$\Delta_f G°$	Log K_f
	J K⁻¹mol⁻¹			kJ mol⁻¹			
0	0.	0.	INFINITE	-8.671	-113.805	-113.805	INFINITE
100	29.104	165.850	223.530	-5.769	-112.415	-120.239	62.807
200	29.108	186.025	200.317	-2.855	-111.286	-128.526	33.568
298.15	29.142	197.653	197.653	0.	-110.527	-137.163	24.030
300	29.142	197.833	197.653	0.054	-110.516	-137.328	23.911
400	29.342	206.238	198.798	2.976	-110.102	-146.338	19.110
500	29.794	212.831	200.968	5.931	-110.003	-155.414	16.236
600	30.443	218.319	203.415	8.942	-110.150	-164.486	14.320
700	31.171	223.066	205.890	12.023	-110.469	-173.518	12.948
800	31.899	227.277	208.305	15.177	-110.905	-182.497	11.916
900	32.577	231.074	210.628	18.401	-111.418	-191.416	11.109
1000	33.183	234.538	212.848	21.690	-111.983	-200.275	10.461
1100	33.710	237.726	214.967	25.035	-112.586	-209.075	9.926
1200	34.175	240.679	216.988	28.430	-113.217	-217.819	9.481
1300	34.572	243.431	218.917	31.868	-113.870	-226.508	9.101
1400	34.920	246.006	220.761	35.343	-114.541	-235.148	8.774
1500	35.217	248.426	222.526	38.850	-115.229	-243.740	8.488
1600	35.480	250.707	224.216	42.385	-115.933	-252.284	8.236
1700	35.710	252.865	225.830	45.945	-116.651	-260.784	8.013
1800	35.911	254.912	227.396	49.526	-117.364	-269.242	7.813
1900	36.091	256.859	228.897	53.126	-118.133	-277.658	7.633
2000	36.250	258.714	230.342	56.744	-118.896	-286.034	7.470
2100	36.392	260.486	231.736	60.376	-119.675	-294.373	7.322
2200	36.518	262.181	233.081	64.021	-120.470	-302.672	7.186
2300	36.636	263.809	234.382	67.683	-121.278	-310.936	7.062
2400	36.321	265.360	235.641	71.324	-122.133	-319.164	6.946
2500	36.836	266.854	236.860	74.985	-122.964	-327.356	6.840
2600	36.934	268.300	238.041	78.673	-133.854	-335.514	6.741
2700	37.003	269.695	239.188	82.369	-134.731	-343.638	6.648
2800	37.083	271.042	240.302	86.074	-135.623	-351.728	6.562
2900	37.150	272.345	241.384	89.786	-136.533	-359.789	6.480
3000	37.217	273.606	242.437	93.504	-137.457	-367.818	6.404
3100	37.279	274.827	243.463	97.327	-138.397	-375.813	6.332
3200	37.338	276.011	244.461	100.960	-129.353	-383.778	6.265
3300	37.392	277.161	245.435	104.696	-130.325	-391.714	6.200
3400	37.443	278.278	246.385	108.438	-131.313	-399.620	6.139
3500	37.493	279.364	247.311	112.185	-132.313	-407.497	6.082
3600	37.543	280.421	248.216	115.937	-133.329	-415.345	6.027
3700	37.589	281.450	249.101	119.693	-134.360	-423.165	5.974
3800	37.631	282.453	249.966	123.454	-135.406	-430.956	5.924
3900	37.673	283.431	250.811	127.219	-136.464	-438.720	5.876
4000	37.715	284.386	251.638	130.986	-137.537	-446.457	5.830
4100	37.756	285.317	252.448	134.763	-138.623	-454.166	5.786
4200	37.794	286.228	253.242	138.540	-139.723	-461.849	5.744
4300	37.833	287.117	254.020	142.331	-140.836	-469.506	5.703
4400	37.869	287.986	254.782	146.106	-141.963	-477.138	5.664
4500	37.903	288.839	255.530	149.895	-143.103	-484.741	5.627
4600	37.941	289.673	256.262	153.687	-144.257	-492.321	5.590
4700	37.974	290.489	256.982	157.483	-145.424	-499.875	5.555
4800	38.007	291.289	257.688	161.282	-146.605	-507.404	5.522
4900	38.041	292.073	258.382	165.084	-147.800	-514.909	5.489
5000	38.074	292.842	259.064	168.890	-149.000	-522.387	5.457
5100	38.104	293.596	259.733	172.699	-150.231	-529.843	5.427
5200	38.137	294.336	260.392	176.511	-151.469	-537.275	5.397
5300	38.171	295.063	261.039	180.326	-152.721	-544.681	5.368
5400	38.200	295.777	261.676	184.144	-153.987	-552.065	5.340
5500	38.074	296.476	262.302	187.957	-155.279	-559.426	5.313
5600	38.263	297.164	262.919	191.775	-156.585	-566.763	5.287
5700	38.296	297.843	263.526	195.603	-157.899	-574.075	5.261
5800	38.325	298.506	264.123	199.434	-159.230	-581.364	5.236
5900	38.335	299.163	264.711	203.268	-160.579	-588.631	5.211
6000	38.388	299.808	265.291	207.106	-161.945	-595.875	5.186

PREVIOUS: September 1965 (1 atm)   CURRENT: September 1965 (1 bar)

Carbon Monoxide (CO) $C_1O_1(g)$

Carboe Diexide (CO_2) IDEAL GAS

W = 44.0098

	Enthalpy Reference Temperature = T_r = 298.15 K			Standard State Pressure = $p°$ = 0.1 MPa			
		J K⁻¹mol⁻¹			kJ mol⁻¹		
T/K	$C_p°$	S°	$-[G°-H°(T_r)]/T$	$H°-H°(T_r)$	$\Delta_f H°$	$\Delta_f G°$	Log K_f
0	0.	0.	INFINITE	-9.364	-393.151	-393.151	INFINITE
100	29.208	179.009	243.568	-6.456	-393.208	-393.683	205.639
200	32.359	199.975	217.046	-3.414	-393.404	-394.085	102.924
298.15	37.129	213.795	213.795	0.	-393.522	-394.389	69.095
300	37.221	214.025	213.795	0.069	-393.523	-394.394	68.670
400	41.325	225.314	215.307	4.003	-393.583	-394.675	51.539
500	44.627	234.901	218.290	8.305	-393.666	-394.939	41.256
600	47.321	243.283	221.772	12.907	-393.803	-395.182	34.404
700	49.564	250.750	225.388	17.754	-393.983	-395.398	29.505
800	51.434	257.494	228.686	22.806	-394.188	-395.586	25.829
900	52.999	263.645	232.500	28.030	-394.405	-395.748	22.969
1000	54.308	269.299	235.901	33.397	-394.623	-395.886	20.679
1100	55.409	274.528	239.176	38.884	-394.838	-396.001	18.805
1200	56.342	279.390	242.329	44.473	-395.050	-396.098	17.242
1300	57.137	283.932	245.356	50.148	-395.257	-396.177	15.919
1400	57.802	288.191	248.265	55.896	-395.462	-396.240	14.784
1500	58.379	292.199	251.062	61.705	-395.668	-396.288	13.800
1600	58.886	295.983	253.753	67.568	-395.876	-396.323	12.939
1700	59.317	299.566	256.343	73.480	-396.090	-396.344	12.178
1800	59.701	302.968	258.840	79.431	-396.311	-396.353	11.502
1900	60.049	306.205	261.248	85.419	-396.542	-396.349	10.896
2000	60.350	309.293	263.574	91.438	-396.784	-396.333	10.351
2100	60.622	312.244	265.822	97.488	-397.039	-396.304	9.858
2200	60.865	315.070	267.996	103.563	-397.309	-396.262	9.408
2300	61.086	317.781	270.102	109.660	-397.596	-396.206	8.998
2400	61.287	320.385	272.144	115.778	-397.900	-396.142	8.623
2500	61.471	332.890	274.124	121.917	-398.222	-396.062	8.275
2600	61.647	325.305	276.046	128.073	-398.562	-395.969	7.955
2700	61.802	327.634	277.914	134.246	-398.921	-395.862	7.658
2800	61.952	329.885	279.730	140.433	-399.299	-395.742	7.383
2900	62.095	332.061	281.497	146.636	-399.695	-395.609	7.126
3000	62.229	334.169	283.218	152.852	-400.111	-395.461	6.886
3100	62.347	336.211	284.895	159.081	-400.545	-395.298	6.661
3200	62.462	338.192	286.529	165.321	-400.998	-395.122	6.450
3300	62.573	340.116	288.124	171.573	-401.470	-394.932	6.251
3400	62.681	341.986	289.681	177.836	-401.960	-394.726	6.064
3500	62.785	343.804	291.202	184.109	-402.467	-394.506	5.888
3600	62.884	345.574	292.687	190.393	-402.991	-394.271	5.731
3700	62.980	347.299	294.140	196.686	-403.532	-394.032	5.563
3800	63.074	348.979	295.561	202.989	-404.089	-393.756	5.413
3900	63.166	350.619	296.952	209.301	-404.662	-393.477	5.270
4000	63.254	352.218	298.314	215.622	-405.251	-393.183	5.134
4100	63.341	353.782	299.648	221.951	-405.856	-392.874	5.005
4200	63.426	355.310	300.955	228.290	-406.475	-392.550	4.882
4300	63.509	356.803	302.236	234.637	-407.110	-392.210	4.764
4400	63.588	358.264	303.493	240.991	-407.760	-391.857	4.652
4500	63.667	359.694	304.726	247.354	-408.426	-391.488	4.544
4600	63.745	361.094	305.937	253.725	-409.106	-391.105	4.441
4700	63.823	362.466	307.125	260.103	-409.802	-390.700	4.342
4800	63.893	363.810	308.292	266.489	-410.514	-390.292	4.247
4900	63.968	365.128	309.438	272.882	-411.242	-389.863	4.156
5000	64.046	366.422	310.565	279.283	-411.986	-389.419	4.068
5100	64.128	367.691	311.673	285.691	-412.746	-388.959	3.984
5200	64.230	368.937	312.762	292.109	-413.522	-388.486	3.902
5300	64.312	370.161	313.833	298.535	-414.314	-387.996	3.824
5400	64.404	371.364	314.888	304.971	-415.123	-387.493	3.748
5500	64.486	372.547	315.925	311.416	-415.949	-386.974	3.675
5600	64.588	373.709	316.947	317.870	-416.794	-386.439	3.605
5700	64.680	374.853	317.953	324.334	-417.658	-385.890	3.536
5800	64.772	375.979	318.944	330.806	-418.541	-385.324	3.470
5900	64.865	377.087	319.920	337.288	-419.445	-384.745	3.406
6000	64.957	378.178	320.882	343.779	-420.372	-384.148	3.344

PREVIOUS: September 1965 (1 atm)   CURRENT: September 1965 (1 bar)

Carboe Diexide (CO_2) C_1O_2(g)

Fluorine (F$_2$)   REFERENCE STATE   W = 37.996806

Enthalpy Reference Temperature = T$_r$ = 298.15 K   Standard State Pressure = p° = 0.1 MPa

T/K	C$_p$	S°	-[G°-H°(T$_r$)]/T	H°-H°(T$_r$)	Δ$_f$H°	Δ$_f$G°	Log K$_f$
		J K^{-1} mol^{-1}			kJ mol^{-1}		
0	0.	0.	INFINITE	-8.825	0.	0.	0.
100	29.114	170.370	229.548	-5.918	0.	0.	0.
200	29.665	190.652	205.594	-2.988	0.	0.	0.
250	30.447	197.354	203.300	-1.456	0.	0.	0.
298.15	31.302	202.788	202.788	0.	0.	0.	0.
300	31.336	202.983	202.790	0.058	0.	0.	0.
350	32.207	207.880	203.173	1.647	0.	0.	0.
400	32.992	212.233	204.040	3.277	0.	0.	0.
450	33.674	216.150	205.172	4.844	0.	0.	0.
500	34.286	219.738	206.452	6.643	0.	0.	0.
600	35.166	226.068	209.268	10.116	0.	0.	0.
700	35.832	231.542	212.016	13.668	0.	0.	0.
800	36.336	236.361	214.764	17.277	0.	0.	0.
900	36.732	240.664	217.407	20.932	0.	0.	0.
1000	37.057	244.552	219.930	24.622	0.	0.	0.
1100	37.334	248.097	222.332	28.341	0.	0.	0.
1200	37.579	251.356	224.617	32.087	0.	0.	0.
1300	37.802	254.373	226.791	35.857	0.	0.	0.
1400	38.008	257.182	228.863	39.647	0.	0.	0.
1500	38.199	259.811	230.838	43.458	0.	0.	0.
1600	38.374	262.282	232.728	47.287	0.	0.	0.
1700	38.530	264.613	234.536	51.132	0.	0.	0.
1800	38.663	266.819	236.268	54.992	0.	0.	0.
1900	38.770	268.913	237.932	58.864	0.	0.	0.
2000	38.846	270.904	239.531	62.745	0.	0.	0.
2100	38.889	272.800	241.071	66.632	0.	0.	0.
2200	38.895	274.609	242.554	70.521	0.	0.	0.
2300	38.864	276.338	243.986	74.409	0.	0.	0.
2400	38.795	277.991	245.369	78.293	0.	0.	0.
2500	38.690	279.572	246.705	82.167	0.	0.	0.
2600	38.549	281.087	247.999	86.030	0.	0.	0.
2700	38.375	282.539	249.251	89.876	0.	0.	0.
2800	38.170	283.931	250.465	93.704	0.	0.	0.
2900	37.938	285.266	251.642	97.509	0.	0.	0.
3000	37.683	286.548	252.785	101.290	0.	0.	0.
3100	37.406	287.779	253.894	105.045	0.	0.	0.
3200	37.111	288.962	254.971	108.771	0.	0.	0.
3300	36.802	290.100	256.019	112.467	0.	0.	0.
3400	36.482	291.193	257.037	116.131	0.	0.	0.
3500	36.152	292.246	258.028	119.763	0.	0.	0.
3600	35.817	293.260	258.993	123.361	0.	0.	0.
3700	35.477	294.237	259.932	126.926	0.	0.	0.
3800	35.135	295.178	260.848	130.457	0.	0.	0.
3900	34.793	296.087	261.740	133.953	0.	0.	0.
4000	34.453	296.963	262.600	137.415	0.	0.	0.
4100	34.114	297.810	263.450	140.844	0.	0.	0.
4200	33.780	298.628	264.285	144.236	0.	0.	0.
4300	33.450	299.419	265.093	147.400	0.	0.	0.
4400	33.125	300.184	265.882	150.929	0.	0.	0.
4500	32.807	300.925	266.653	154.225	0.	0.	0.
4600	32.495	301.642	267.405	157.490	0.	0.	0.
4700	32.191	302.336	268.141	160.734	0.	0.	0.
4800	31.893	303.013	268.861	163.929	0.	0.	0.
4900	31.603	303.667	269.565	167.103	0.	0.	0.
5000	31.331	304.303	270.253	170.349	0.	0.	0.
5100	31.046	304.920	270.927	173.366	0.	0.	0.
5200	30.779	305.521	271.586	176.459	0.	0.	0.
5300	30.518	306.104	272.232	179.524	0.	0.	0.
5400	30.266	306.673	272.865	182.563	0.	0.	0.
5500	30.023	307.226	273.484	185.577	0.	0.	0.
5600	29.787	307.765	274.092	188.568	0.	0.	0.
5700	29.557	308.290	274.687	191.535	0.	0.	0.
5800	29.335	308.803	275.271	194.480	0.	0.	0.
5900	29.119	309.301	275.843	197.402	0.	0.	0.
6000	28.911	309.789	276.405	200.304	0.	0.	0.

Fluorine (F$_2$)   F$_2$(ref)

Fluorine (F) IDEAL GAS

W = 18.998403

Enthalpy Reference Temperature = T_r = 298.15 K   Standard State Pressure = p° = 0.1 MPa

T/K	C_p°	S°	$-[G°-H°(T_r)]/T$	$H°-H°(T_r)$	$\Delta_f H°$	$\Delta_f G°$	Log K_f
		J K⁻¹mol⁻¹			kJ mol⁻¹		
0	0.	0.	INFINITE	-6.518	77.284	77.284	INFINITE
100	21.205	134.479	176.805	-4.433	77.916	73.987	-38.124
200	22.603	149.670	160.833	-2.233	78.652	67.783	-17.703
250	22.783	154.736	158.125	-1.097	79.037	65.021	-13.585
298.15	22.746	158.750	158.750	0.	79.390	62.289	-10.913
300	22.742	158.891	158.751	0.042	79.403	62.183	-10.837
350	23.603	162.387	159.027	1.176	79.743	59.286	-8.848
400	23.431	165.394	159.639	2.302	80.053	56.343	-7.368
450	23.250	168.026	160.428	3.419	80.337	53.361	-6.194
500	23.100	170.363	161.307	4.528	80.597	50.350	-5.260
600	21.832	174.366	163.161	6.724	81.086	44.284	-3.853
700	21.628	177.717	165.006	8.807	81.453	38.090	-2.842
800	21.474	180.594	166.760	11.051	81.803	31.871	-2.081
900	21.357	183.117	168.458	13.193	82.117	25.611	-1.486
1000	21.266	185.362	170.038	15.334	82.403	19.317	-1.009
1100	21.194	187.365	171.525	17.446	82.666	12.998	-0.617
1200	21.137	189.227	172.934	19.563	82.909	6.651	-0.290
1300	21.091	190.917	174.244	21.674	83.136	0.287	-0.012
1400	21.053	192.478	175.492	23.781	83.348	-6.095	0.227
1500	21.022	193.930	176.673	25.885	83.548	-12.490	0.435
1600	20.996	195.286	177.795	27.986	83.733	-18.899	0.617
1700	20.974	196.558	178.861	30.085	83.908	-25.319	0.778
1800	20.955	197.756	179.878	32.181	84.075	-31.748	0.921
1900	20.938	198.889	180.849	34.276	84.234	-38.188	1.050
2000	20.925	199.963	181.778	36.369	84.387	-44.635	1.166
2100	20.913	200.983	182.669	38.461	84.535	-51.090	1.271
2200	20.902	201.956	183.523	40.551	84.681	-57.551	1.366
2300	20.893	202.885	184.345	42.641	84.827	-64.020	1.454
2400	20.885	203.774	185.136	44.730	84.974	-70.494	1.534
2500	20.877	204.626	185.899	46.818	85.125	-76.975	1.606
2600	20.871	205.445	186.635	48.906	85.281	-83.462	1.677
2700	20.865	206.232	187.346	50.993	85.444	-89.956	1.740
2800	20.860	206.991	188.034	53.079	85.617	-96.455	1.799
2900	20.855	207.723	188.701	55.164	85.800	-102.961	1.855
3000	20.851	208.430	189.347	57.250	85.994	-109.473	1.906
3100	20.847	209.114	189.973	59.334	86.202	-115.992	1.954
3200	20.843	209.775	190.582	61.419	86.423	-122.518	2.000
3300	20.840	210.417	191.173	63.503	86.660	-129.051	2.043
3400	20.837	211.039	191.748	65.587	86.911	-135.592	2.083
3500	20.834	211.643	192.308	67.670	87.179	-142.140	2.121
3600	20.831	212.230	192.854	69.754	87.463	-148.696	2.156
3700	20.829	212.800	193.385	71.837	87.764	-155.250	2.192
3800	20.827	213.356	193.903	73.920	88.081	-161.832	2.225
3900	20.825	213.897	194.409	76.002	88.416	-168.413	2.256
4000	20.823	214.424	194.903	78.084	88.767	-175.003	2.285
4100	20.821	214.938	195.385	80.167	89.135	-181.602	2.314
4200	20.820	215.440	195.857	82.349	89.520	-188.209	2.341
4300	20.818	215.930	196.318	84.331	89.921	-194.827	2.367
4400	20.817	216.408	196.769	86.412	90.338	-201.454	2.392
4500	20.815	216.876	197.211	88.494	90.771	-208.090	2.415
4600	20.814	217.334	197.643	90.575	91.220	-214.736	2.438
4700	20.813	217.781	198.067	92.657	91.685	-221.392	2.461
4800	20.812	218.219	198.482	94.738	92.164	-228.059	2.482
4900	20.811	218.648	198.889	96.819	92.658	-234.736	2.502
5000	20.810	219.069	199.289	98.900	93.166	-241.422	2.522
5100	20.809	219.481	199.681	100.981	93.687	-248.119	2.541
5200	20.808	219.885	200.065	103.062	94.223	-254.826	2.560
5300	20.807	220.281	200.443	105.143	94.771	-261.544	2.579
5400	20.807	220.670	200.814	107.224	95.332	-268.272	2.595
5500	20.806	221.052	201.179	109.304	95.906	-275.011	2.612
5600	20.805	221.427	201.537	111.385	96.491	-281.760	2.628
5700	20.805	221.795	201.889	113.465	97.088	-288.530	2.644
5800	20.804	222.157	202.235	115.546	97.696	-295.290	2.659
5900	20.803	222.513	202.576	117.626	98.315	-302.071	2.674
6000	20.803	222.862	202.911	119.706	98.944	-308.862	2.689

PREVIOUS: September 1965 (1 atm)   CURRENT: June 1982 (1 bar)

Fluorine (F) F_1(g)

INDEX

INDEX

Printed in the United States
by Baker & Taylor Publisher Services